GW01458905

/1500
£7.50

ATLAS OF DINOFLAGELLATES

Oxytoxum reticulatum (Stein) Schütt
from the East Atlantic. Cell 45 μm long.

ATLAS OF DINOFLAGELLATES

A Scanning Electron Microscope Survey

JOHN D. DODGE

Botany Department,
Royal Holloway & Bedford Colleges, (University of London)
Egham, Surrey

1985

FARRAND PRESS
LONDON

FARRAND PRESS
50 FERRY STREET
ISLE OF DOGS
LONDON E14

ISBN: 1 85083 004 5

British Library Cataloguing in Publication Data
Dodge, John D.
    Atlas of dinoflagellates.
    1. Dinoflagellata
    I. Title
    593.1'8   QL368.D6
    ISBN 1–85083–004–5

Typeset in Ehrhardt and printed by Latimer Trend & Company Ltd
on Parilux Gloss paper
Bound by the Standard Bookbinding Company Ltd.

CONTENTS

This book is dedicated to
Janet
for all her encouragement,
help, and love.

Preface

Dinoflagellates, or red-tide organisms as they are often known, form one of the most important components of plankton. They are small single-celled organisms, some have an animal-like nutrition, others possess chloroplasts and, like land plants, are primary producers of food. Both types form important links in aquatic food webs. Some dinoflagellates make their presence known by bioluminescence which can light up the sea at night, particularly in the tropics. Others are a scourge of shell-fish farmers since they contain toxin (Paralytic Shellfish Poison or PSP), which can affect man after it has been concentrated by filter feeding molluscs. Yet others, such as the Florida red-tide organism, produce toxins which can kill fish. Clearly, these organisms have an economic importance quite out of keeping with their microscopic size.

Many present-day dinoflagellates have a resistant stage, or cyst, in their life-cycle which is generally produced at the onset of adverse conditions. These cysts provide an important taxonomic link with fossilised dinocysts which are present in abundance in sedimentary deposits at least as far back as Jurassic times (over 150 million years). Knowledge of how and where cysts form today can clearly be of assistance in interpreting the stratigraphic record of fossil cysts.

Both in terms of numbers of species and abundance of cells, dinoflagellates are very much more common in the sea than in freshwater lakes and ponds. This fact is reflected in the present book where only 5 per cent of the illustrations are of freshwater species.

Dinoflagellates, in spite of being only single-celled organisms, are very varied in their morphology, as a quick glance at the plates will show. Some are extremely beautiful, others are bizarre, whilst some are positively grotesque. These aesthetic qualities are often reflected in their Latin names! In this book the great advantages of the Scanning Electron Microscope—depth of focus, ability to view the specimen from various angles—are utilised to add valuable detail to what can be seen by light microscopy. Clearly, the SEM has proved to be of considerable value in the taxonomy of dinoflagellates since it enables us to see more clearly characters visible by light microscopy and also adds finer details of the cell wall structures. At the same time we are able to appreciate the varied forms of these minute creatures and enjoy them as art forms which may give ideas or inspiration for quite unrelated activities. I hope that these micrographs will open up new vistas for many, besides providing a reference work for marine biologists and palynologists. In the course of our researches over the past six years we have amassed a collection of some 25 thousand scanning electron micrographs of dinoflagellates. For this book I have tried to cover as many genera and species as possible whilst using only clear and informative micrographs.

February 1985   JOHN D. DODGE

Introduction

What are Dinoflagellates?

In simple terms dinoflagellates are small (usually less than 100 μm) single-celled organisms which swim freely in water. They are propelled by two flagella which in general are orientated one around the cell, within a groove termed the girdle, and one directed to the posterior. The flagella are not shown in any of the pictures in this book since they easily detach from the cell when it dies or is badly preserved. In any case, to obtain a clear picture of the thecal plates (or armour), it is advantageous to have lost the membranes around the outside of the cell and these also include the flagellar membrane. This book is concerned exclusively with species which have this armoured plating but it should be noted that there are hundreds of so called naked species which are bounded only by a membranous covering. These organisms are very easily damaged and are difficult to study by any form of microscopy. In addition, there are several dinoflagellates whose main life stage is as a parasite or as a symbiont associated with an invertebrate host. These are not included in the present book.

The name 'dino' comes from a Greek word meaning 'whirling' as this type of movement is characteristic of these organisms. Dinoflagellates have several unique subcellular structures which set them apart from the other groups which belong to the algae and Protozoa. These include the cell covering, chloroplasts, trichocysts, the pusule, and particularly the mesocaryotic nucleus (see Dodge, 1983; Spector, 1984, for details).

Many dinoflagellates are known to have a resistant stage, usually termed a cyst, or dinocyst. This has a very tough wall which is either impregnated with virtually indestructible sporopollenin or may be covered with minerals such as calcium carbonate (as in *Scrippsiella*) or silica (as in *Ceratium hirundinella*: see Chapman, Dodge and Heaney, 1982). Cysts settle to the bottom of the sea or lake and when conditions are suitable they may germinate, the new motile stage emerging through an excystment aperture or archeopyle. Studies of germination in the laboratory have enabled us to link up many of the often elaborately ornamented cysts with their motile counterparts (e.g. Wall and Dale, 1968; Lewis, Dodge and Tett, 1984). Large numbers of cysts are known from the fossil record, but only a selection of modern cysts is included in this book. For further details of cysts see Evitt, 1985.

Ecology and distribution

Dinoflagellates are of necessity aquatic organisms. Perhaps the largest number of species is planktonic, swimming freely, or floating, in the sea or in large bodies of water. They are found in all latitudes from the Arctic and Antarctic seas to the tropics and it is in the warm waters that they are largest in size and most bizarre in form. Those photographed for this book mainly come from north temperate regions, but a few tropical forms are included. At present so little is known about the nutrition and physiology of dinoflagellates that we can only guess that the great extension of the cell in forms like *Ornithocercus* is to provide a greater surface area for nutrient uptake or for buoyancy.

Many dinoflagellates occur in marginal habitats. Some are confined to near-shore or neritic waters which are probably more nutritious and of lower salinity than the open ocean water. Several are to be found living amongst the sand-grains of beaches (see Saunders and Dodge, 1984) whilst others are found in the extremely changeable habitat provided by the pools on rocky shores and salt marshes. In freshwater locations some species inhabit small and relatively ephemeral pools whilst others are to be found only in large lakes and reservoirs.

The theca (or amphiesma)

In the taxonomy of dinoflagellates the cell covering, or theca, provides the most useful descriptive characters. In the first place it is clearly responsible for the shape of the cell and it is this which gives the broad classification. Secondly, in armoured dinoflagellates the main part of the theca consists of cellulosic plates, and it is the number, shape, arrangement, and ornamentation of these which provide features to make possible the finer taxonomic separation.

The main characteristics of the theca are illustrated in Fig. 1. There are usually two distinct portions, the upper or *epitheca* and lower or *hypotheca*. Here they are shown as more or less equal in size, but the proportions can vary enormously. The epi- and hypotheca are separated by a transverse groove called the *girdle* (or cingulum). This is, in fact, the location of the transverse flagellum and it may run straight across or be displaced, to a greater or lesser extent, as in this figure. There is generally also a longitudinal groove or furrow which is termed the *sulcus*. This relates to the point of insertion and orientation of the longitudinal flagellum.

The anterior end of the cell is called the *apex* and here there may be a structure of unknown function termed the *apical pore*. At the opposite end of the cell, or *antapex*, there are often projecting spines or horns. The side of a cell in which the flagella insertion can be seen is termed *ventral* and the opposite side (or back) is *dorsal*. When referring to the left or right side of the cell these are as seen from the cell; i.e. the cell's

2

APEX

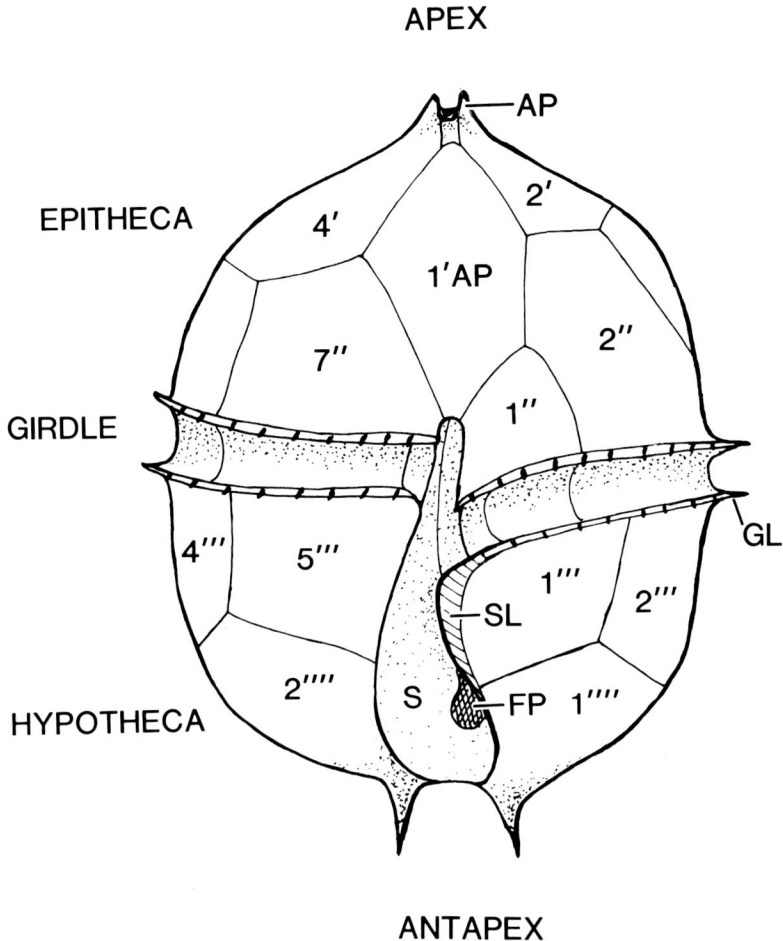

FIG. 1.   A diagrammatic ventral view of a peridinioid dinoflagellate to illustrate some of the terms used in the legends to the micrographs. The numbering system for thecal plates is also shown in part where n' represents the apical series; n'' the precingular series; n''' the postcingular series; and n'''' the antapical plates. AP: apical pore, FP: flagellar pore, GL: girdle list, S: sulcus, SL: sulcal list, l'AP: first apical plate.

left side is on the right of the picture (when the cell is seen in ventral view, as in Fig. 1).

In defining the thecal plates and also naming them to describe their shape or points of contact, a coding system is often used which was first invented by Kofoid. For this it is assumed that there are basically four series of plates around a peridinoid cell such as that shown in Fig. 1. There is an apical series commencing with the distinctive first apical (1' AP) plate which touches both the sulcus and the apical pore. In the genus

Protoperidinium the shape of the first apical plate is an important taxonomic character (4-sided = ortho; 5-sided = meta; 6-sided = para). The second series is that anterior to the girdle (1''). The third is that posterior to the girdle (1'''), and the final series comprises the antapicals (1''''). In some organisms there are extra plates which have to be termed either anterior (a) or posterior (p) intercalaries. There are also plates to be numbered in both the girdle (c) and sulcus (s), although these are often much more difficult to determine. Thus, the Kofoidean tabulation formula for the hypothetical cell shown in Fig. 1 would be:

$$Po, 4', 7'', 3c, 3s, 5''', 2''''$$

Formulae will be given in some of the legends. The junctions between plates are termed sutures and these may be noteworthy because of their ornamentation or extension which may be related to the growth of the cell.

Advantages of Scanning Electron Microscopy (SEM)

SEM is an ideal tool for observing and photographing dinocysts and the wall or theca of motile dinoflagellates. In the first place, when the correct accelerating voltage is used, the narrow beam of electrons is designed to strike the outside of the cell and reflect from it; there is no confusion as a result of seeing into or through the cell. Secondly, the SEM has a great depth of focus—about 300 times that of the light microscope. As a result of the varied contrast over the specimen, the image can have a really three-dimensional appearance. Thirdly, the modern SEM has a very good resolving power. On the instrument used for the micrographs in this book the resolution is around 7 nm, which is a great improvement over the 200 nm which can be obtained on a good light microscope. In addition, the SEM has a very good range of magnifications, $15 \times -100\,000 \times$ which are readily obtained by turning a switch. The fourth advantage of the SEM for this type of work is that the specimen can be both rotated and tilted. Thus, when an interesting cell has been discovered it can generally be examined and photographed from almost every angle, except the side which is adhering to the specimen support stub.

Collection and preparation methods for dinoflagellates

Many samples were collected from the seashores around much of the British Isles, whilst others were obtained from the East Atlantic, North Sea, etc. generally by colleagues from other laboratories carrying out marine biological or oceanographic researches. Some samples were obtained from freshwater lakes and ponds. Mostly, the samples were concentrated using filters, nets, or sedimentation methods and were

4

preserved in formaldehyde or Lugol's iodine (see Sournia, 1978, for methods). In the laboratory the material was first examined by light microscopy and the dinoflagellates present identified.

For the best results with SEM individual cells were picked out of the samples by means of a micropipette (using a binocular low-power microscope). They were washed in freshwater to remove salt and preservatives and were then re-pipetted onto a Nuclepore filter membrane (8 μm pores) held in a special block. The holes seen in the background to many of the photographs are the pores in the supporting filter membrane. When several cells had been collected on the filter, a graded series of acetone concentrations was passed through in order to remove all water. After dehydration the block containing the filters was placed in the chamber of a Polaron critical point drying apparatus (Type E3000) and dried using liquid carbon dioxide. Next the filters were glued to 13-mm diameter stubs (Cambridge pattern) using acetate glue (UHU). They were then coated with a complete layer of gold/palladium approx 50 nm thick using a Polaron 'cool' sputter coater (Type E5100).

All the photographs in this book were taken on a JSM-25S scanning electron microscope (JEOL, UK Ltd, Colindale, London NW9) generally operated at an accelerating voltage of 15 kV. Photographs were taken on Ilford FP4 or HP5 film using a high resolution cathode ray tube. Prints and enlargements were made on Ilford Multigrade photographic paper.

A note on the micrographs and legends

As far as possible a standard ventral view is given for each cell illustrated. Sometimes this is supplemented by a detail or a small picture of a cyst stage. The legend has been kept quite brief. It aims to point out any unusual feature of the organism and to emphasise points of taxonomic importance. Some idea of the distribution/ecology is given and the actual location from which the specimen shown was obtained is stated. The size of the specimen is given in terms of diameter (d), or length (l) and width (w). All pictures are of marine organisms unless indicated in the last line by FW (freshwater).

Outline classification of the free-living dinoflagellates

After Dodge, in: Spector, 1984. Groups in [brackets] are not covered in this book.

Phylum or Division: PYRROPHYTA (or DINOPHYTA)
Class: DINOPHYEAE   *Genera illustrated*
                                             in this Atlas
                                             by one or more species

Order: Prorocentrales
    Family: Prorocentraceae   *Prorocentrum*
                                             Mesoporos

Order: Dinophysiales
    Family: Amphisoleniaceae   *Amphisolenia*
    Family: Dinophysiaceae   *Dinophysis*
    Family: Ornithocercaceae   *Ornithocercus*
[Order: Gymnodiniales]
[Order: Noctilucales]
[Order: Lophodiniales]
Order: Peridinales
    [Family: Pyrophacaceae]
    [Family: Ostreopsidaceae]
    [Family: Crypthecodiniaceae]
    Family: Peridiniaceae   *Boreadinium*
                                             Diplopsalis
                                             Heterocapsa
                                             Oblea
                                             Peridiniopsis
                                             Peridinium
                                             Protoperidinium
                                             Scrippsiella
    Family Gonyaulacaceae   *Gonyaulax*
                                             Protogonyaulax
                                             Peridiniella
                                             Amphidoma
                                             Protoceratium
                                             Pyrodinium
    Family: Ceratocoraceae   *Ceratocorys*
    Family: Triadiniaceae   *Triadinium*
    Family: Heterodiniaceae   *Heterodinium*
    Family: Ceratiaceae   *Ceratium*
    Family: Amphidiniopsidaceae   *Amphidiniopsis*
                                             Planodinium
    Family: Thecadiniaceae   *Thecadinium*
    Family: Oxytoxaceae   *Oxytoxum*
                                             Centrodinium
    Family: Cladopyxidaceae   *Palaeophalocroma*
                                             Cladopyxis
    Family: Podolampaceae   *Podolampas*

6

Each organism illustrated is given a generic name, as above, and its own specific name. This latter is qualified by the 'authority' which is the name of the person or persons responsible for the name which is used. The pictures in this book should not be used on their own in order to identify dinoflagellates since so few species are represented here. A more comprehensive taxonomic work should always be consulted, e.g. Schiller (1933–37), Huber-Pestalozzi (1950), Drebes (1974), Dodge (1982), Taylor (1976).

Acknowledgements
The author is indebted to the Natural Environment Research Council who supported the research projects which enabled the pictures in this book to be taken. Grateful thanks are due to those who assisted with the research and took most of the micrographs: Richard Saunders, as a result of whose enthusiasm this book was started; Jane Lewis, who isolated and photographed most of the cysts; Heather Hermes who developed the techniques; and Debbie Chapman for *Ceratium hirundinella*. Also, to members of staff of the Botany Department for their skilled photography (David Ward and Lynne Etherington) and word-processing (Clare Freeman) on this book. Publication of this book has been aided by support from JEOL, Polaron and Ilford. Finally, a word of thanks for the many researchers who have supplied me with plankton samples and without whose generous assistance this book would scarcely have been possible.

Further Reading

Chapman, D. V., Dodge, J. D. and Heaney, S. I. (1982). Cyst formation in the dinoflagellate *Ceratium hirundinella*. *J. Phycol.* 18, 121–129.

Dodge, J. D. (1975). The Prorocentrales (Dinophyceae) II Revision of the genus *Prorocentrum*. *Bot. J. Linn. Soc.* 71, 103–125.

Dodge, J. D. (1982). Marine Dinoflagellates of the British Isles. Her Majesty's Stationery Office, London.

Dodge, J. D. (1983). Dinoflagellates: Investigation and Phylogenetic Speculation. *Br. Phycol. J.* 18, 335–356.

Dodge, J. D. and Hermes, H. (1981). A revision of the *Diplopsalis* group of dinoflagellates based on material from the British Isles. *Bot. J. Linn. Soc.* 83, 15–26.

Dodge, J. D. and Saunders, R. D. (1985). An SEM study of *Amphidoma nucula* (Dinophyceae) and description of the thecal plates in *A. caudata*. *Arch. Protistenk.* 129, 89–99.

Dodge, J. D. and Saunders, R. D. (1985). A partial revision of the genus *Oxytoxum* (Dinophyceae) with the aid of scanning electron microscopy. *Botanica Marina* 26, 99–122.

Drebes, G. (1974). "Marines Phytoplankton". Thieme, Stuttgart.

Evitt, W. R. (1985). "Sropollenin Dinoflagellate Cysts." American Assn. Stratigraphic Palynologists.

Huber-Pestalozzi, G. (1950). "Das Phytoplankton des Süsswassers". 3 Teil: Cryptophyceen, Chloromonadophyceen, Peridineen. Stuttgart.

Lee, J. J., Hutner, S. H. and Bovee, E. C. (1985). "Illustrated Guide to the Protozoa." Society of Protozoologists, USA.

Lewis, J., Dodge, J. D. and Tett, P. (1984). Cyst-theca relationships in some *Protoperidinium* species (Peridiniales) from Scottish Sea Lochs. *J. Micropalaeontol.* **3**, 25–34.

Saunders, R. D. and Dodge, J. D. (1984). An SEM study and taxonomic revision of some armoured sand-dwelling dinoflagellates. *Protistologica* **20**, 271–283.

Schiller, J. (1933, 1937). "Dinoflagellatae" Vol. 10 pt. 3 sects. I, II. *In*: "Rabenhorst's Kryptogamenflora". Akads Verlag, Leipzig.

Sournia, A. (Ed.) (1978). "Phytoplankton Manual". Unesco, Paris.

Spector, D. (Ed.) (1984). "The Dinoflagellates". Academic Press, Orlando.

Taylor, F. J. R. (1976). Dinoflagellates from the Indian Ocean Expedition. *Bibl. Botanica* **132**, Stuttgart.

Wall, D. and Dale, B. (1968). Modern dinoflagellate cysts and evolution of the Peridiniales. *Micropaleontology* **14**, 265–304.

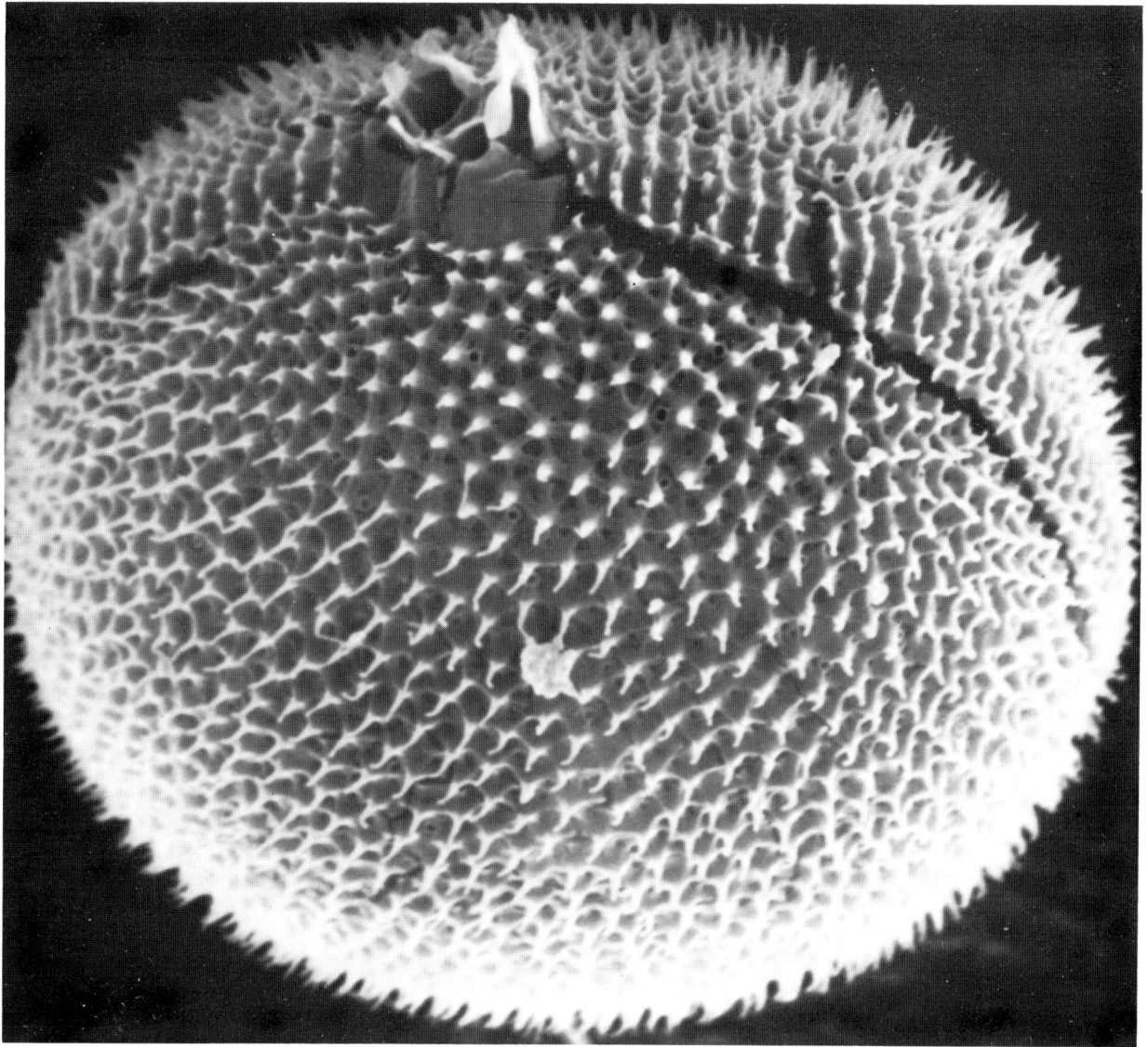

Prorocentrum balticum (Lohmann)
Loeblich
A small, almost spherical, oceanic dinoflagellate with very spiny thecal plates
(detail in small picture). The two flagella
emerge from apical flagellar pores (top)
where there is a very short apical spine.
E. Atlantic. 18 μm d.

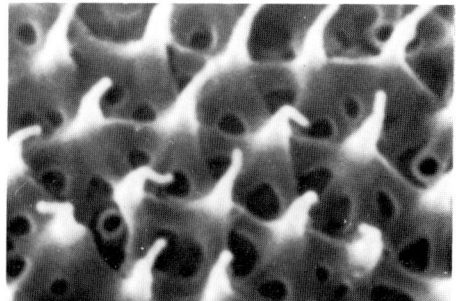

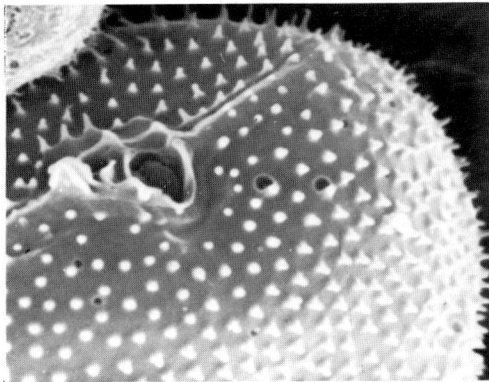

Prorocentrum minimum var _triangulatum_ (Pavillard) Schiller
A small dinoflagellate lacking any appreciable apical spine. The two main thecal plates (only one seen here) are covered with small spines. The flagellar pore area is shown in the small picture. Pagham Harbour, Sussex. 17 μm l, 12 μm w.

10

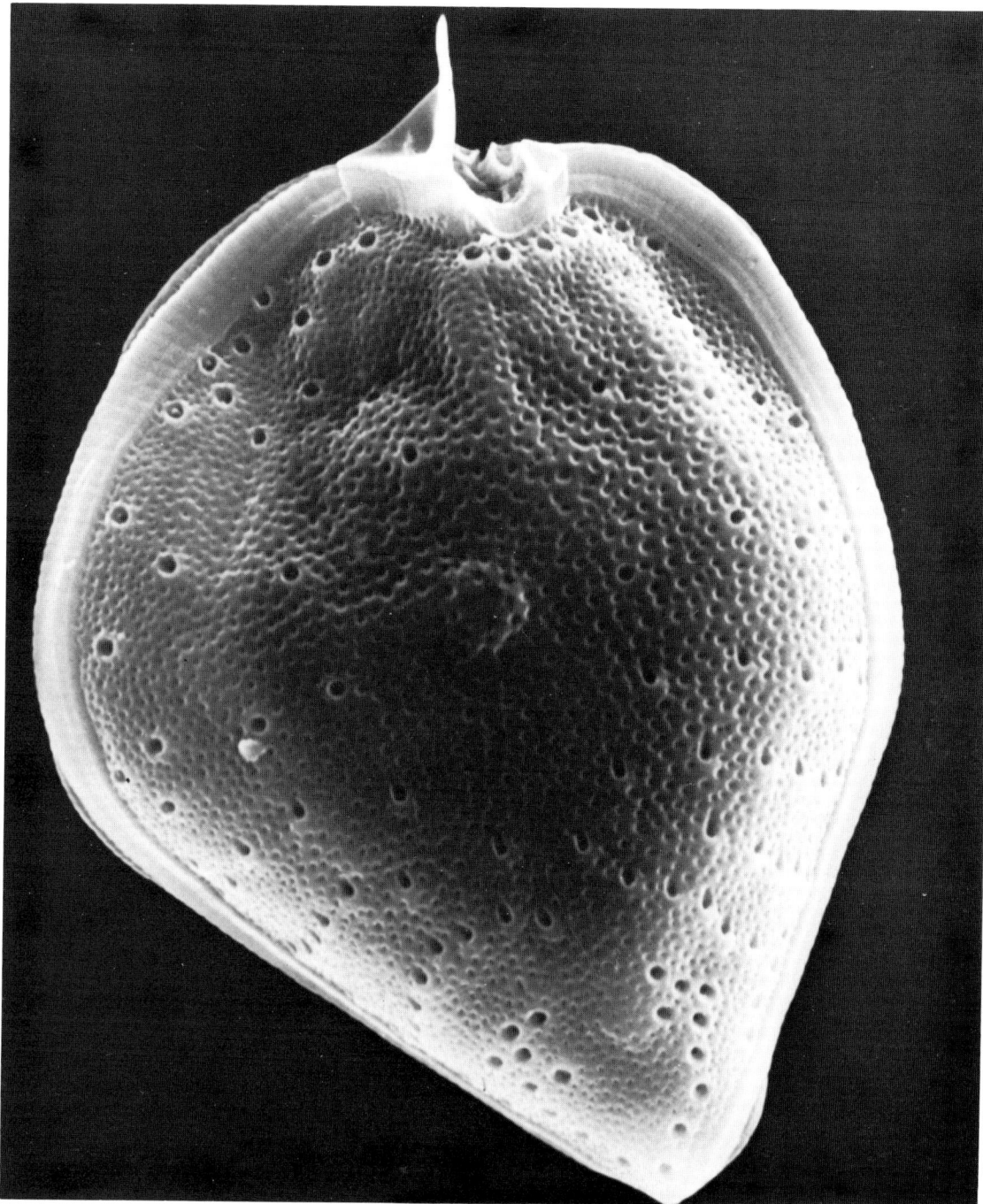

Prorocentrum micans Ehrenberg
A fairly common dinoflagellate of neritic waters. The characteristically flattened theca has
two main plates which are ornamented with small depressions and perforated by trichocyst
pores arranged in a typical pattern. The two flagella emerge through pores in the small
plates beside the apical spine (small picture).
Estuary of River Fal, Cornwall. 45 μm l, 35 μm w.

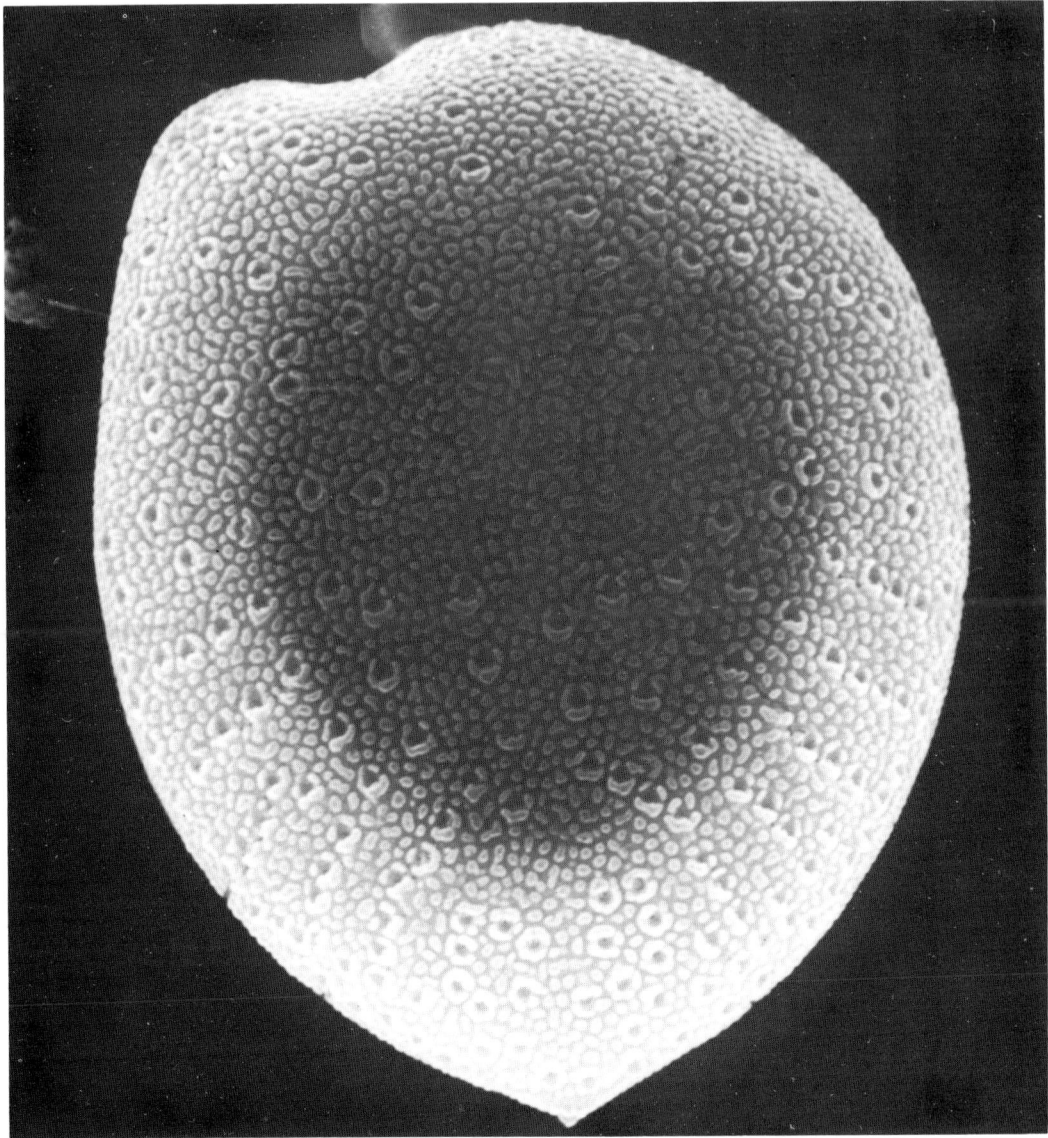

Prorocentrum scutellum Schröder
Rather similar in shape to _P. micans_ and
fairly easily confused when examined by
light microscopy. _P. scutellum_ is more
rounded and has a flatter, less pointed
apical spine. The main thecal plates are
covered with small pimples (small picture) unlike the depressions seen in _P.
micans_. However, the trichocyst pores are
arranged in much the same way. Loch
Creran, Scotland, 50 μm l, 37 μm w.

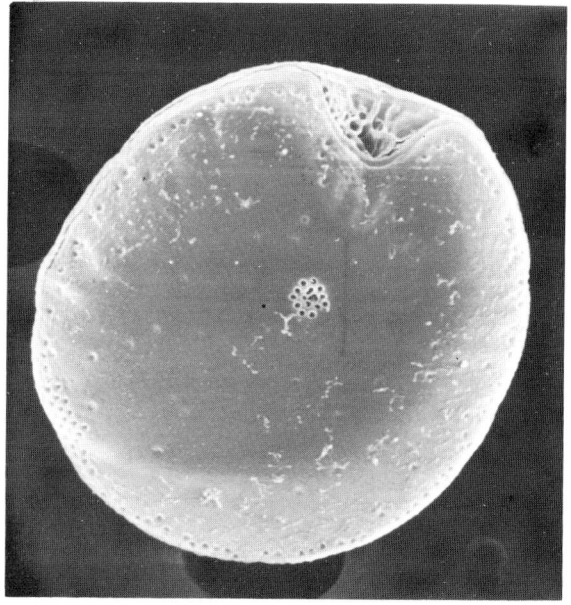
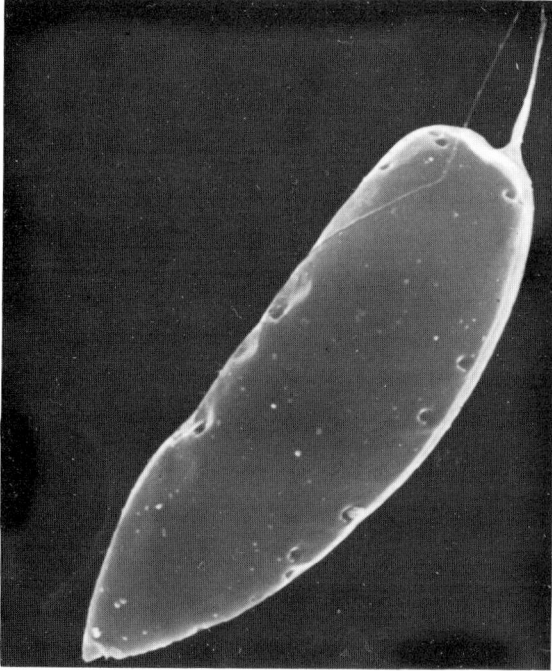
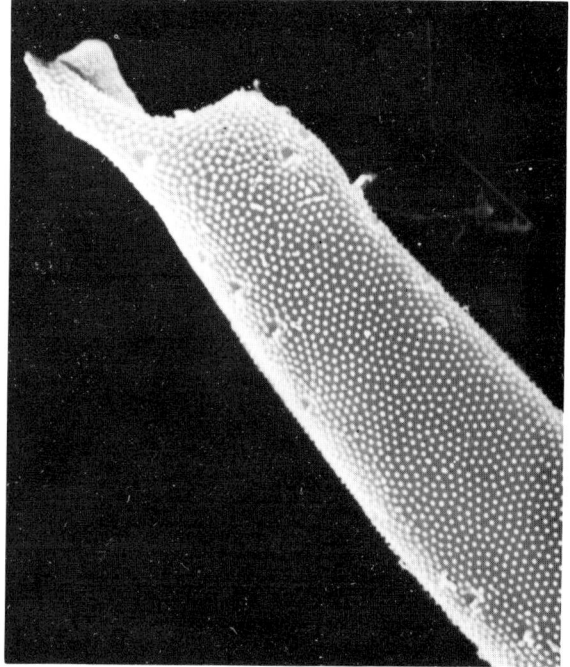

Four species of **Prorocentrum**

A. *P. compressum* Abé: An oceanic species with distinctive thick, patterned thecal plates.
    Atlantic. 30 μm w.
B. *P. lima* (Ehrenberg) Dodge: One form of this smooth-walled beach species. Brittany.
    27 μm w.
C. *P. triestinum* Schiller: A small, delicate oceanic species. E. Atlantic. 30 μm l, 8 μm w.
D. *P. rostratum* Stein: The anterior part of this spiny species. E. Atlantic. 15 μm w.

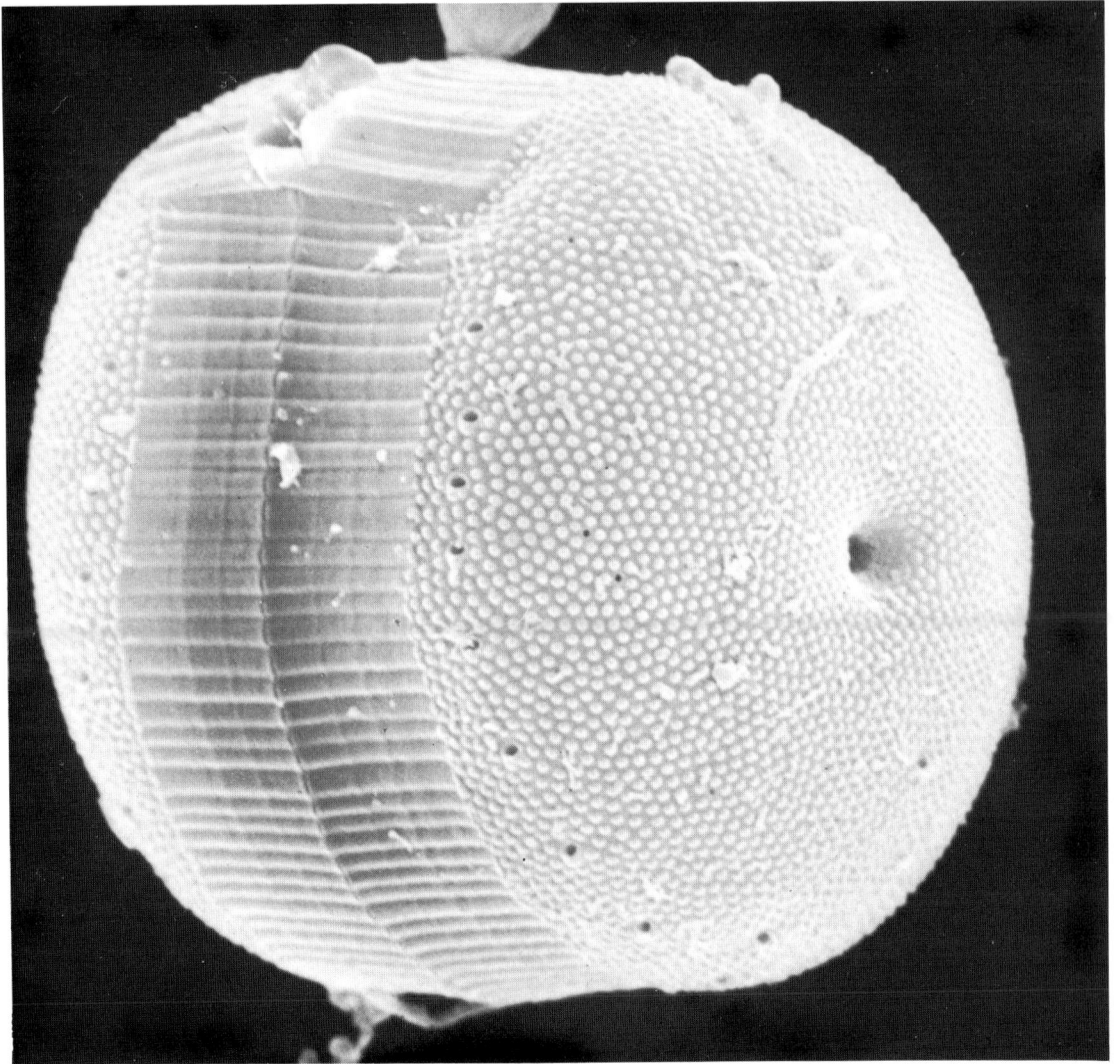

Mesoporos perforatus (Gran) Lillick
This organism is closely related to *Procentrum* but has a
characteristic large pore in the centre of each main thecal plate. The
plates are covered with small pimples and in this specimen have very
broad, ridged, sutural bands around their edges where they join.
E. Atlantic. 21 μm 1, 15 μm w.

14

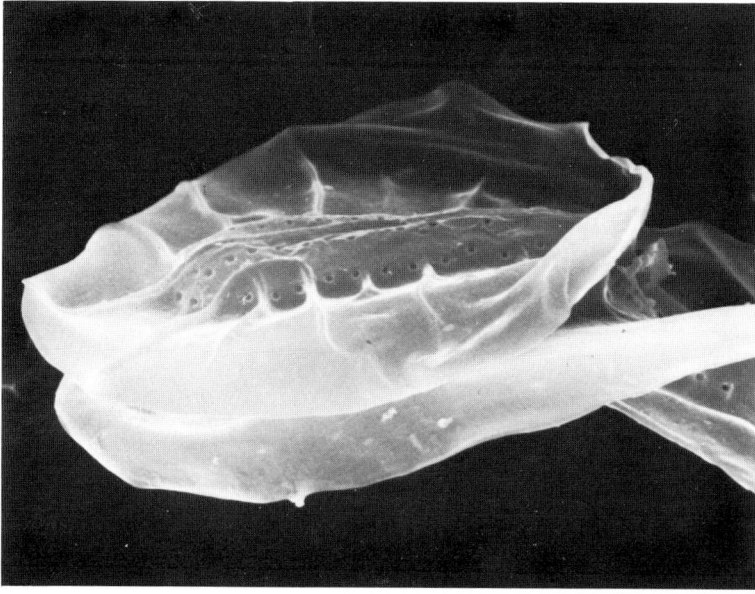
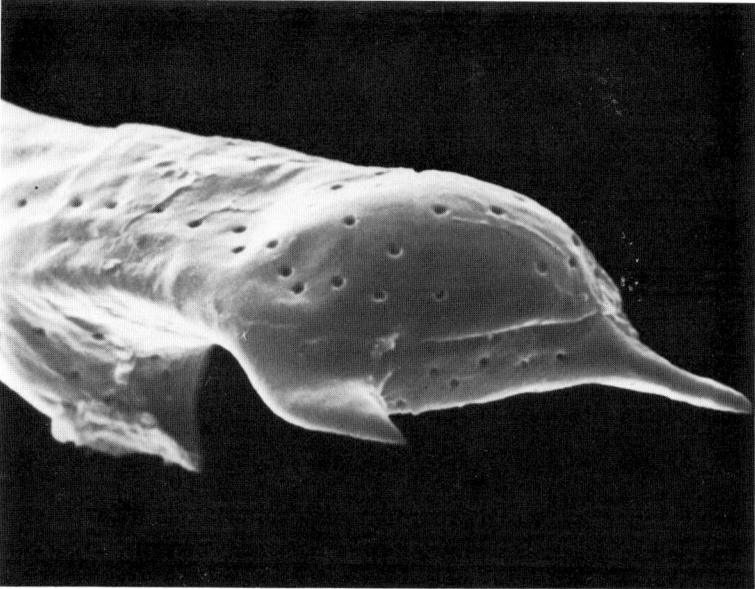
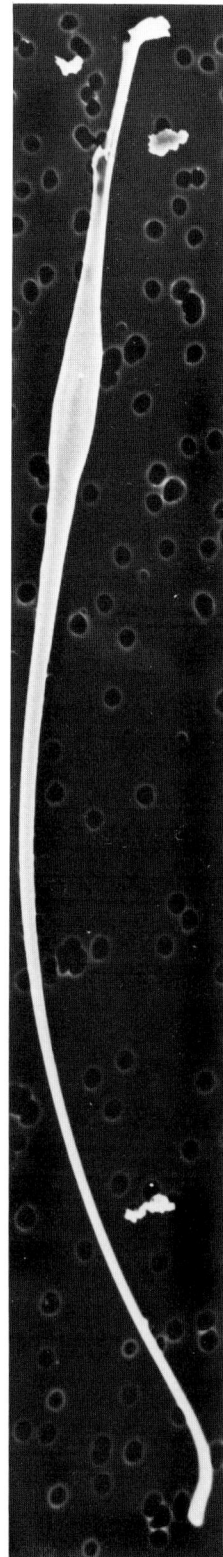

Amphisolenia globifera Stein
An extremely elongated dinoflagellate of
tropical waters. In spite of the odd shape
the thecal plates are organised as in
Dinophysis. Top left: The epitheca and
girdle; lower left: the antapical swelling
at the end of the hypotheca.
E. Mediterranean. 750 μm l.

15

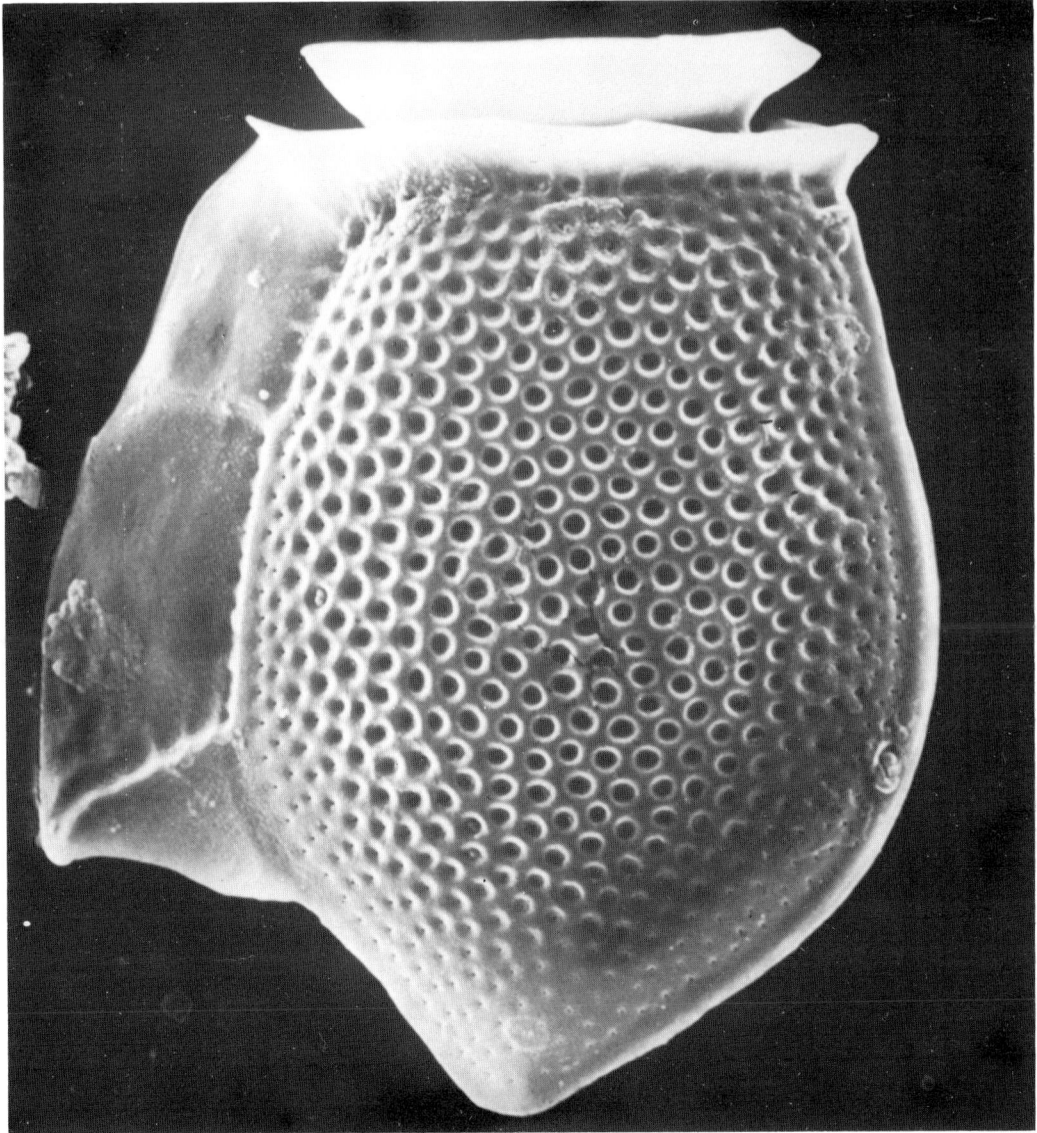

Dinophysis acuta Ehrenberg

A fairly common species from neritic situations around the British Isles. The girdle is virtually at the apex of the cell and the sulcus, with its extensive list, is along one of the narrow sides of the cell. This species is widest about one-third of the length from the antapex.

North Sea. 72 μm l, 55 μm w.

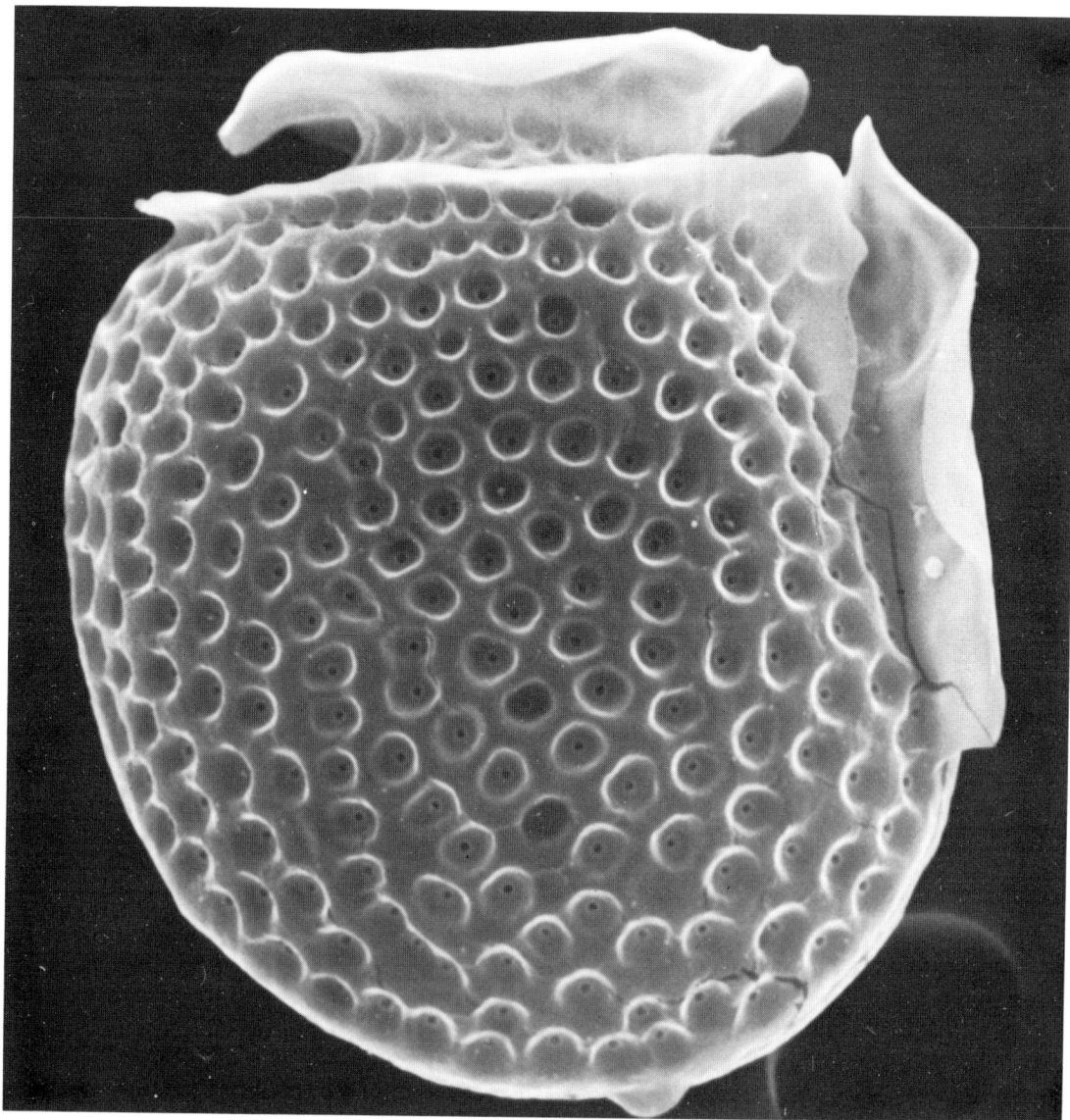

Dinophysis arctica Mereschkowsky
A small species of *Dinophysis* from cool northern waters. The epitheca is reduced to little more that the anterior girdle list. Note the pore at the base of each depression in the hypothecal plate.
Iceland. 35 μm l, 30 μm w.

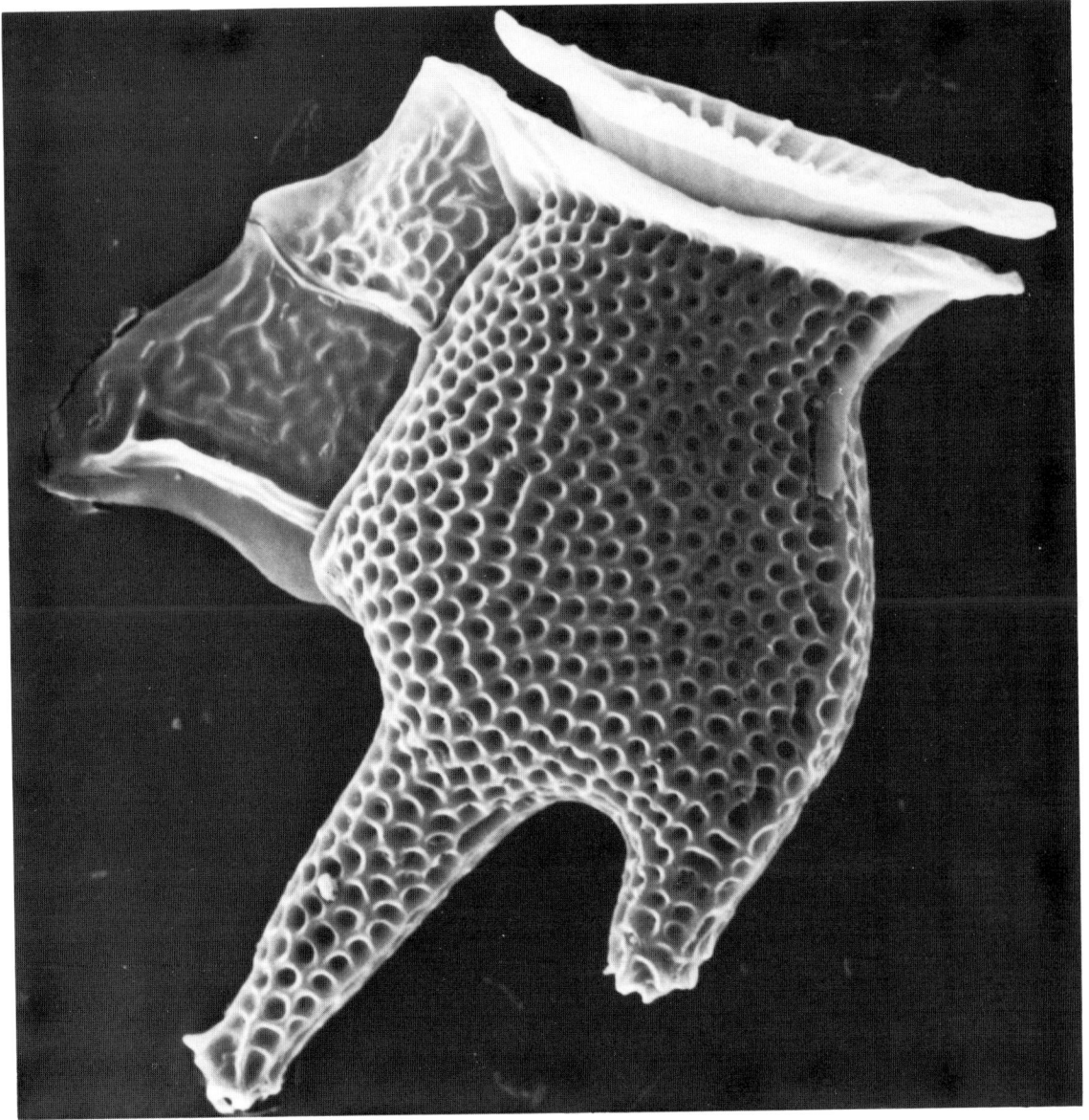

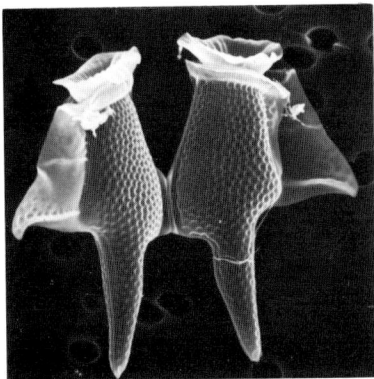

Dinophysis caudata Saville-Kent
A very distinctive _Dinophysis_ which is
often found as a pair of attached cells
(small figure) joined together by the edge of
the hypotheca. Frequently found in the
N.
Atlantic plankton, at times it is brought
near to the British Isles by the Gulf
Stream. Atlantic. 100 μm l, 80 μm w.

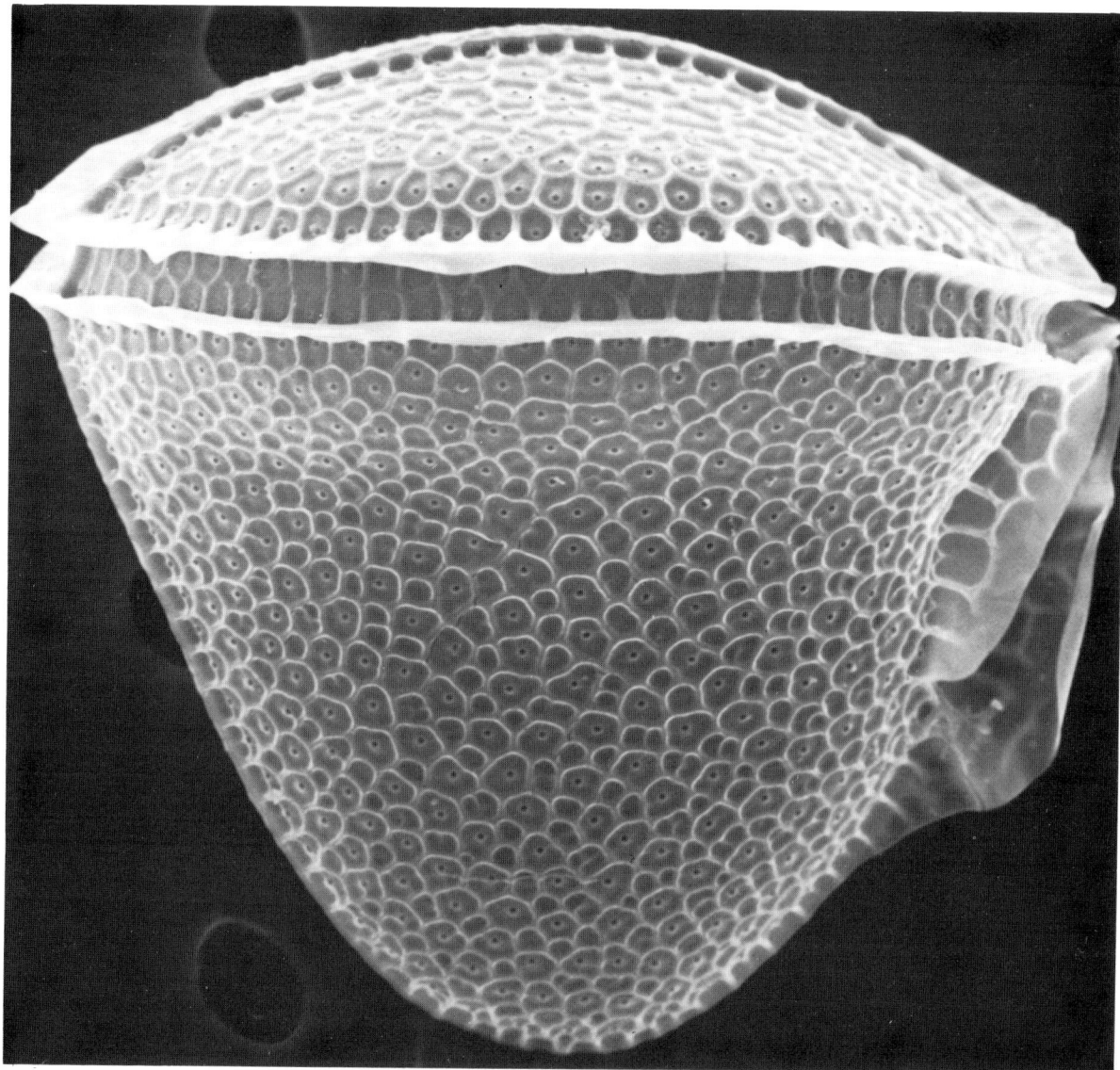

Dinophysis cuneus (Schütt) Abé
An organism from the tropical and
sub-tropical plankton. Here the characteristic
dinophysoid organisation of reduced
epitheca and enlarged hypotheca is
clearly displayed. In this species the
epitheca is domed and the girdle lists are
rather narrow.
E. Atlantic. 75 μm l, 80 μm w.

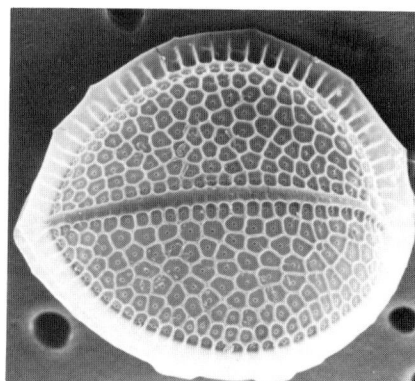

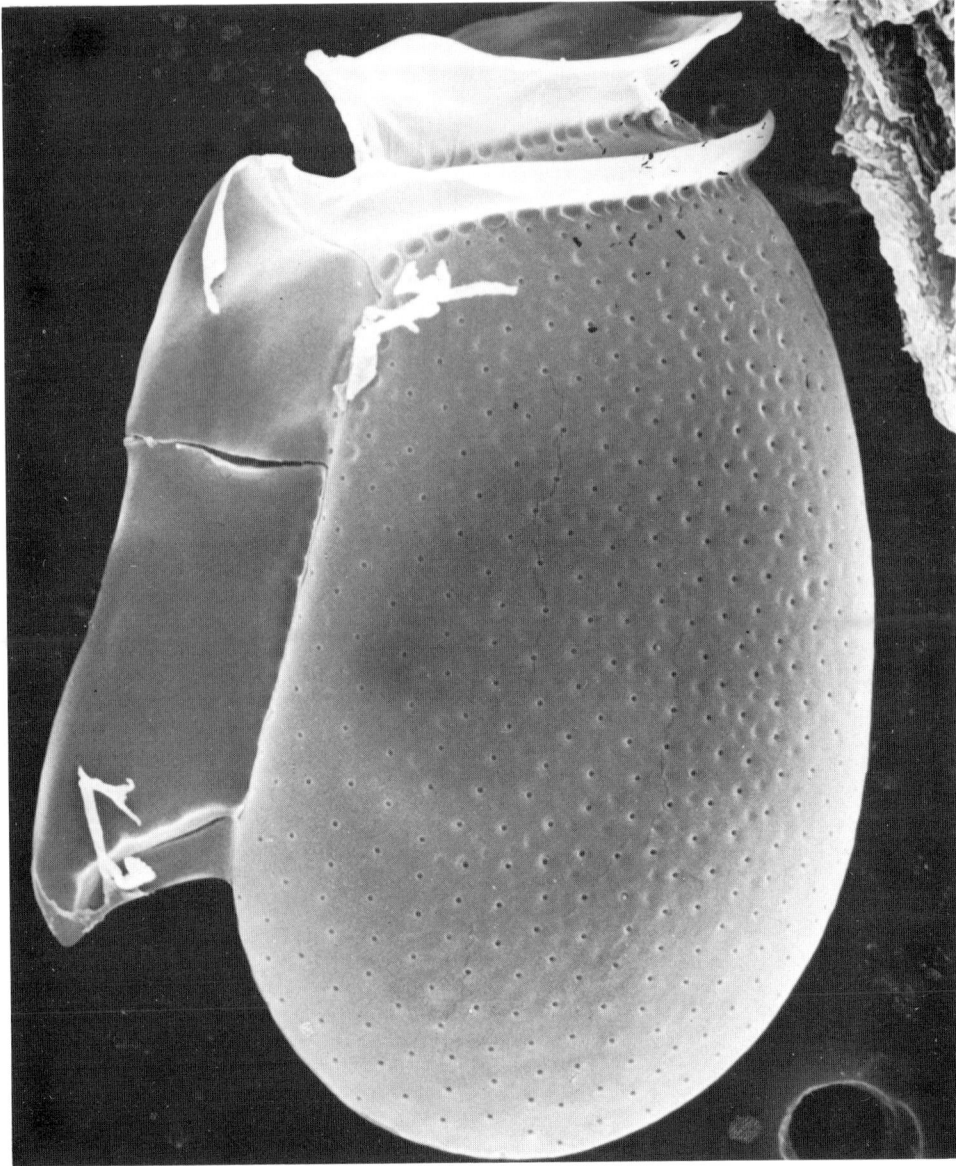

Dinophysis dens Pavillard

A distinctive sac-shaped _Dinophysis_. The main thecal plates are smooth, but, as in all members of this genus, are perforated with numerous small pores. The girdle and sulcal lists are well developed.

E. Atlantic. 56 μm l, 38 μm w.

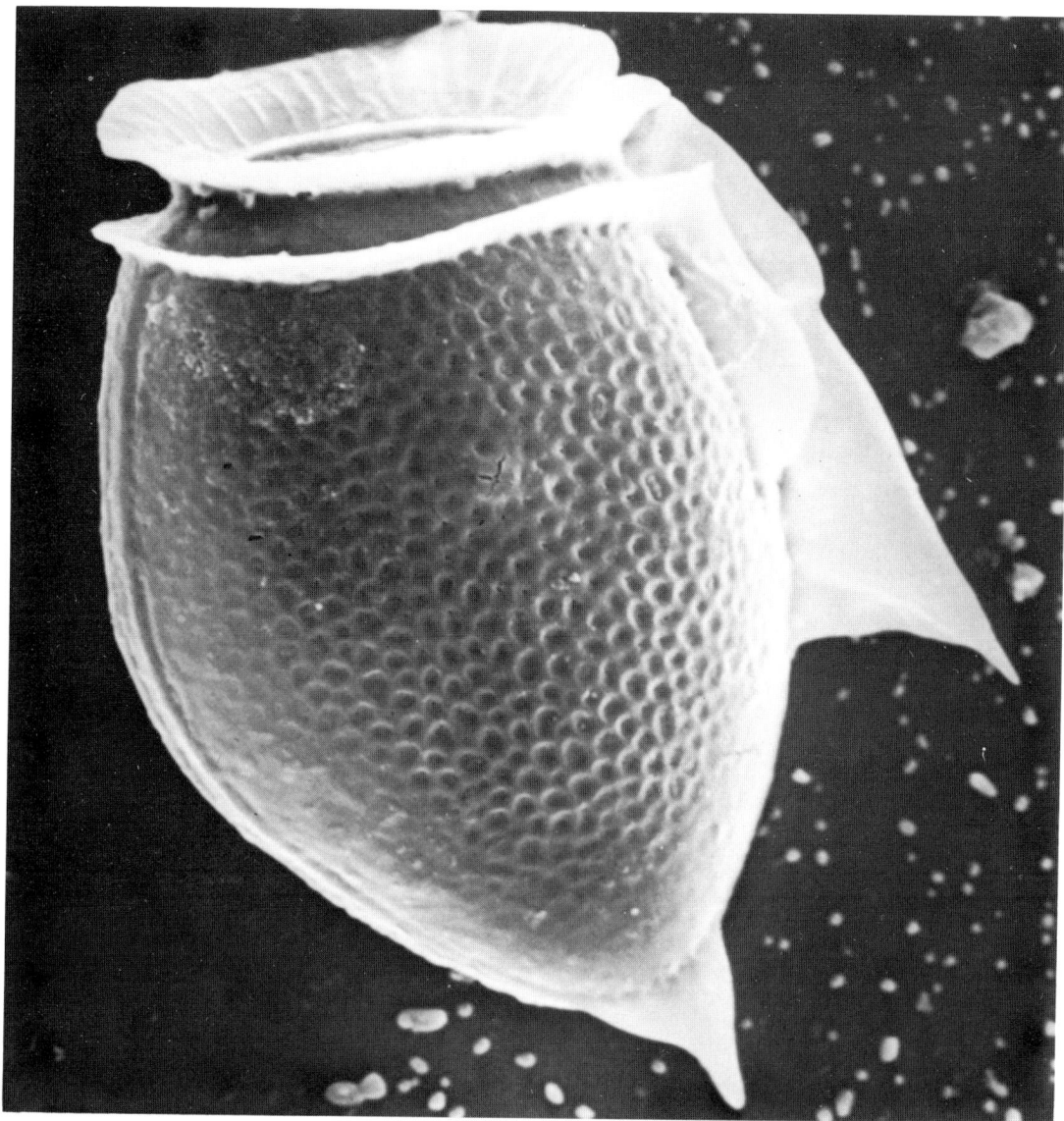

Dinophysis hastata Stein
A large oceanic species with distinctive pointed antapex to the hypotheca and wide pointed left sulcal list. Found in the warm Gulf Stream waters.
E. Atlantic. 72 µm l, 56 µm w.

Dinophysis micropterygia Dangeard
A *Dinophysis* species with a very tiny epitheca. The girdle and sulcal lists are also very narrow but the thecal sculpturing is pronounced even on the sulcal list. A warm–water species. E. Atlantic. 55 μm l, 45 μm w.

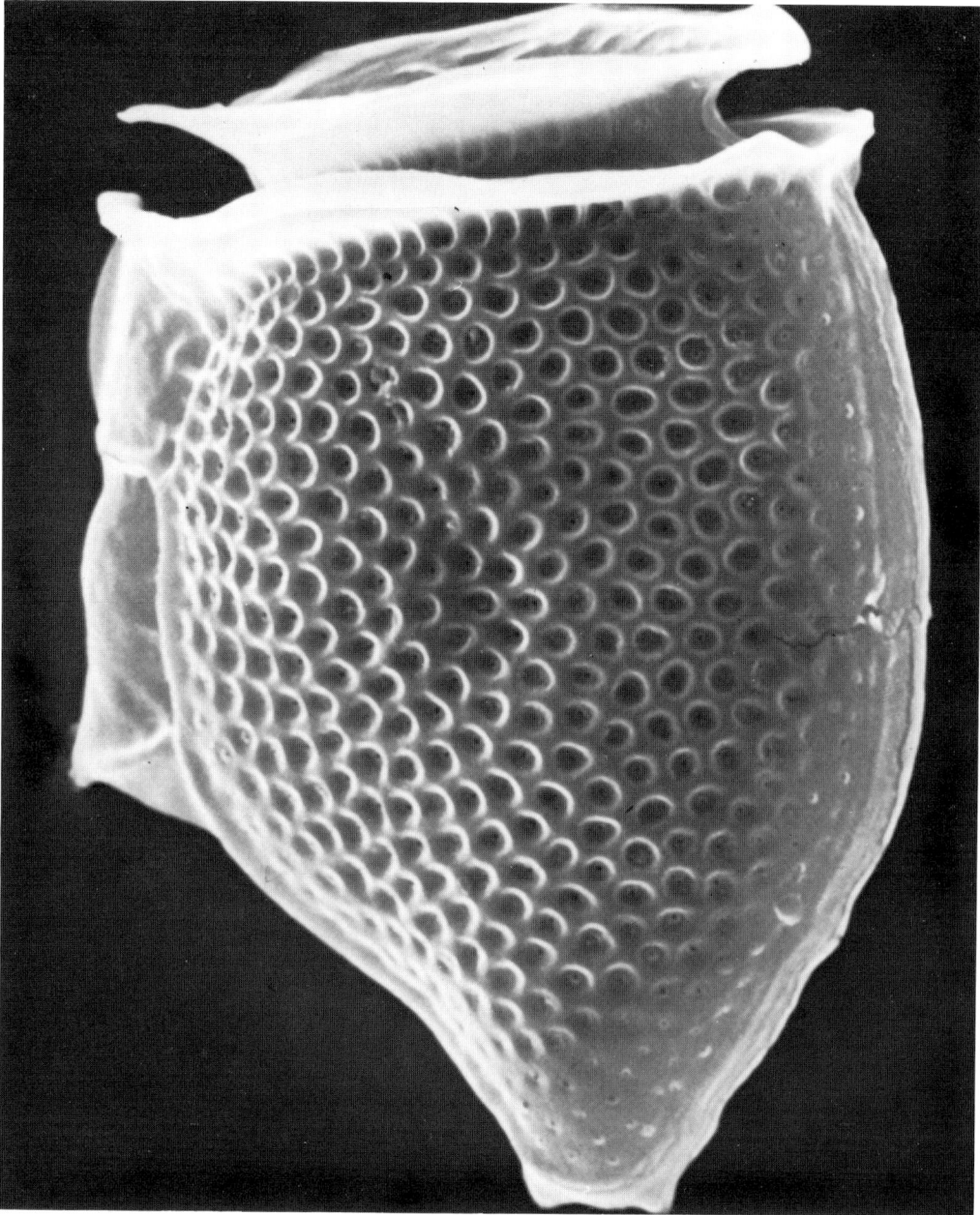

Dinophysis norvegica Claparédè & Lachmann
Rather similar to *D. acuta* but more pointed at the antapex and lacking the hypothecal
bulge. At times *D. norvegica* is common in the North Sea and it is only known from
northern waters.
North Sea, off Whitby. 55 μ l, 40 μm w.

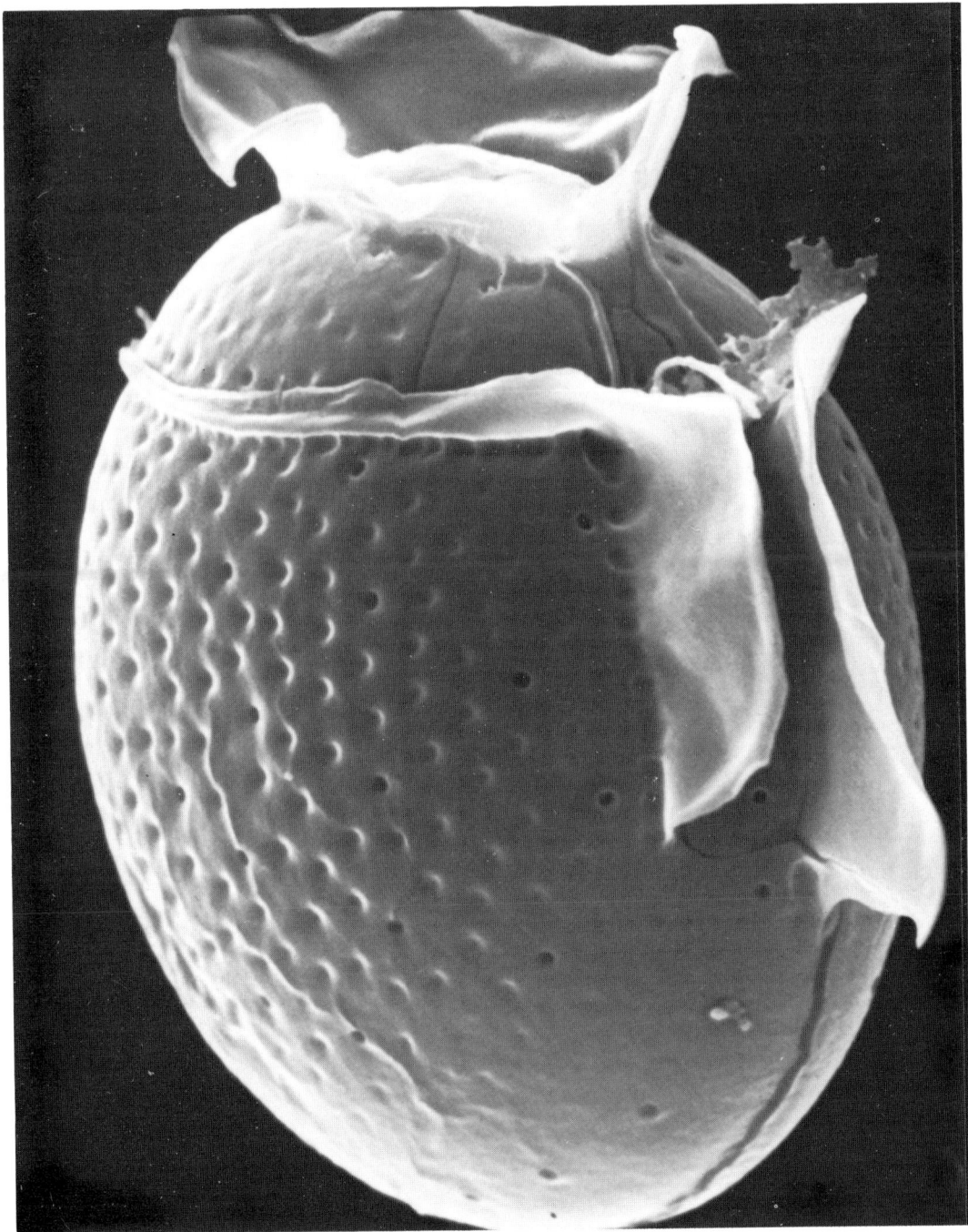

Dinophysis parva Schiller

A distinctly ovoid species with an unusual girdle in which the anterior list is very wide and the posterior list very narrow. The thecal plates are perforated but not heavily sculptured. Note the two sulcal lists, the left being typically larger than the right.

E. Atlantic. 21 µm l, 16 µm w.

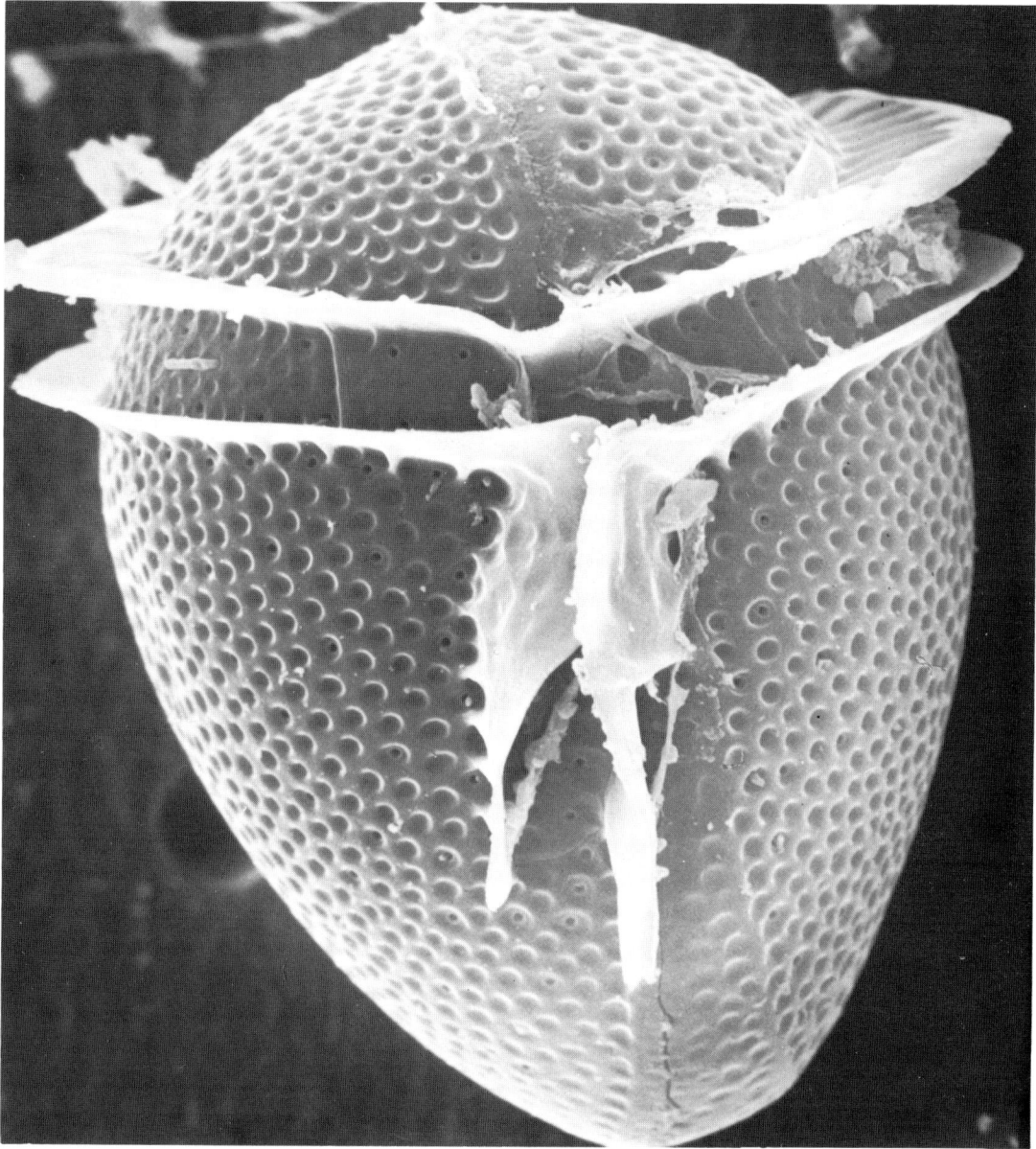

Dinophysis porodictyum (Stein) Abé
A view of the ventral side showing the junction of girdle and sulcus (the point where the flagella emerge). The two main plates of both epitheca and hypotheca can be seen in this picture.
E. Atlantic. 45 μm l, 40 μm w.

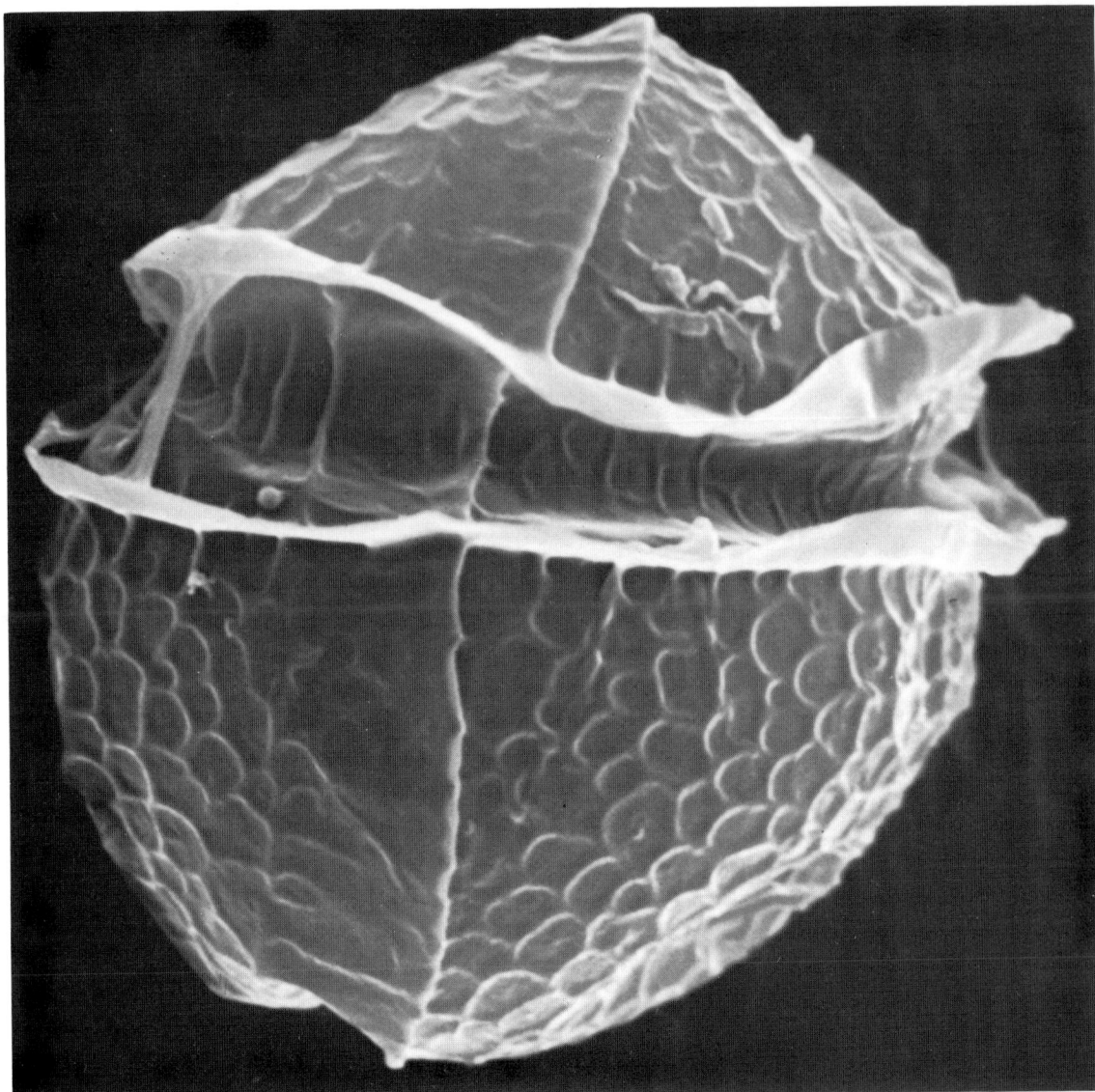

Dinophysis pulchella (Lebour) Balech

A very small species which rather uncharacteristically for this genus is almost as thick as it is long (i.e. nearly spherical). The picture shows a cell from the dorsal side. The girdle is rather wide in relation to the size of the cell.

E. Atlantic. 20 μm l, 18 μm w.

26

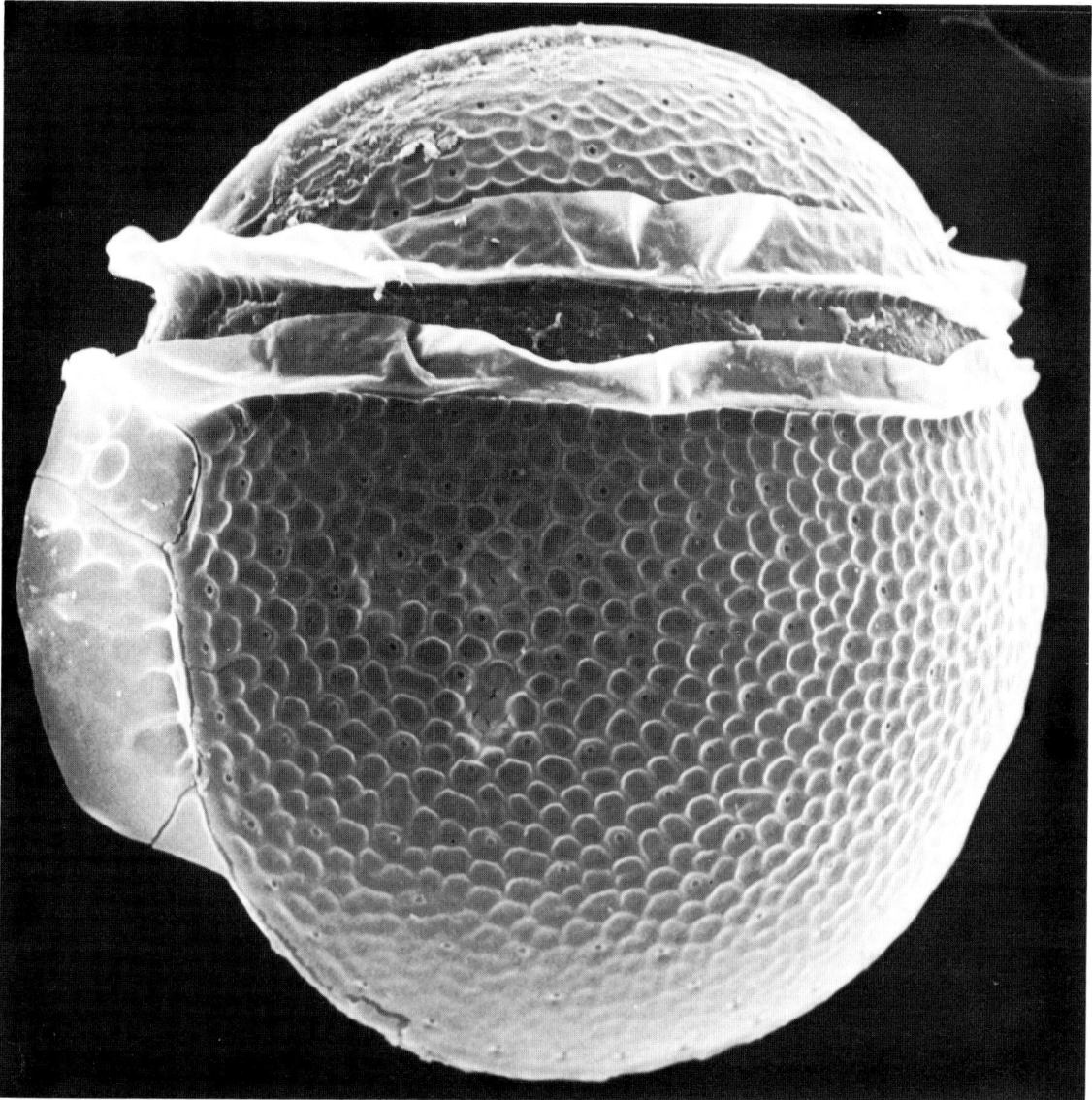

Dinophysis rotundata Claparédè & Lachmann
A fairly common, non-photosynthetic dinoflagellate from neritic waters. Because of its
domed epitheca it is almost round in side view, hence the specific name.
E. Atlantic. 50 μm l, 45 μm w.

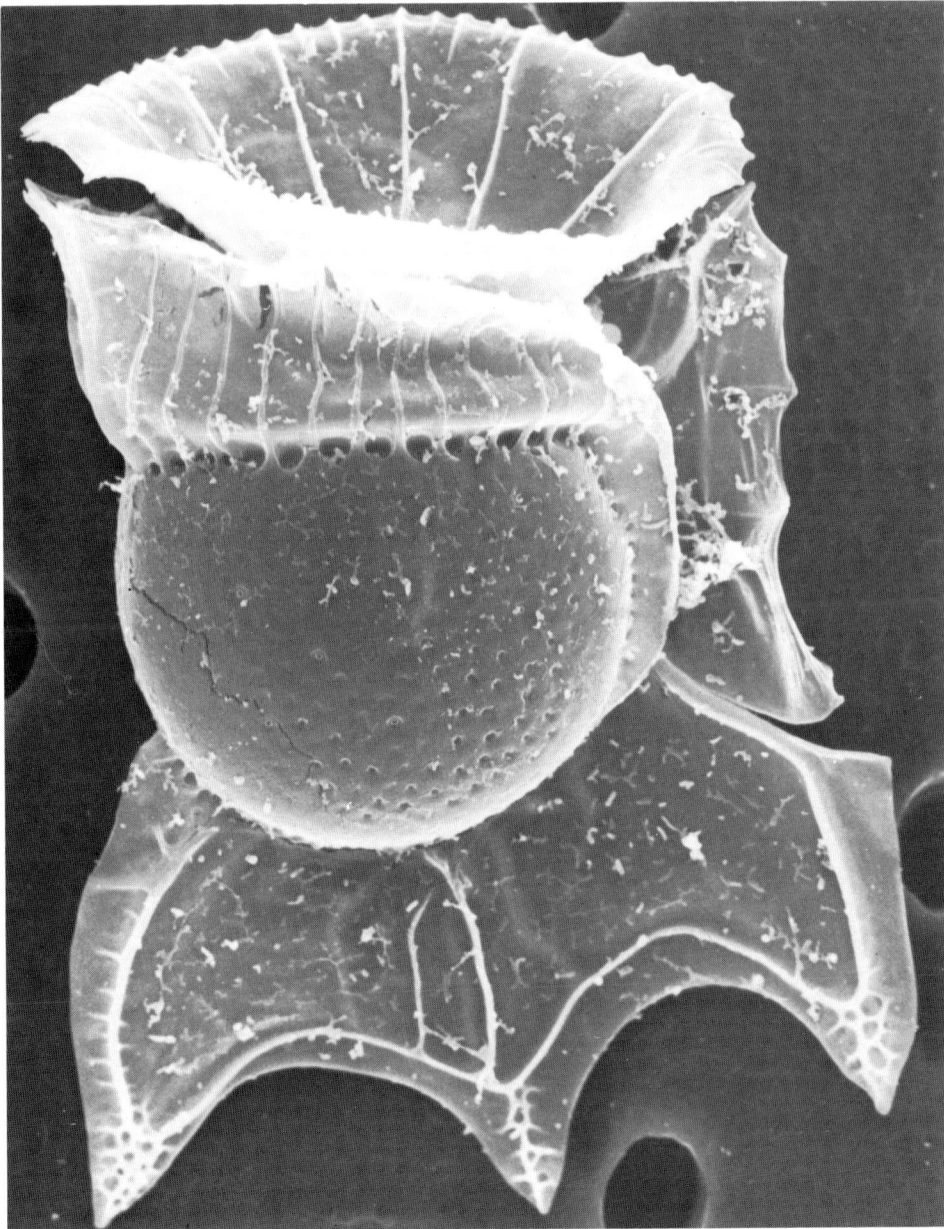

Ornithocercus magnificus Stein
A large and strikingly ornamented species of warm waters. The girdle and sulcal lists are very extensive but much less strongly ridged than in *O. quadratus*.
E. Atlantic. 80 μm l, 60 μm w.

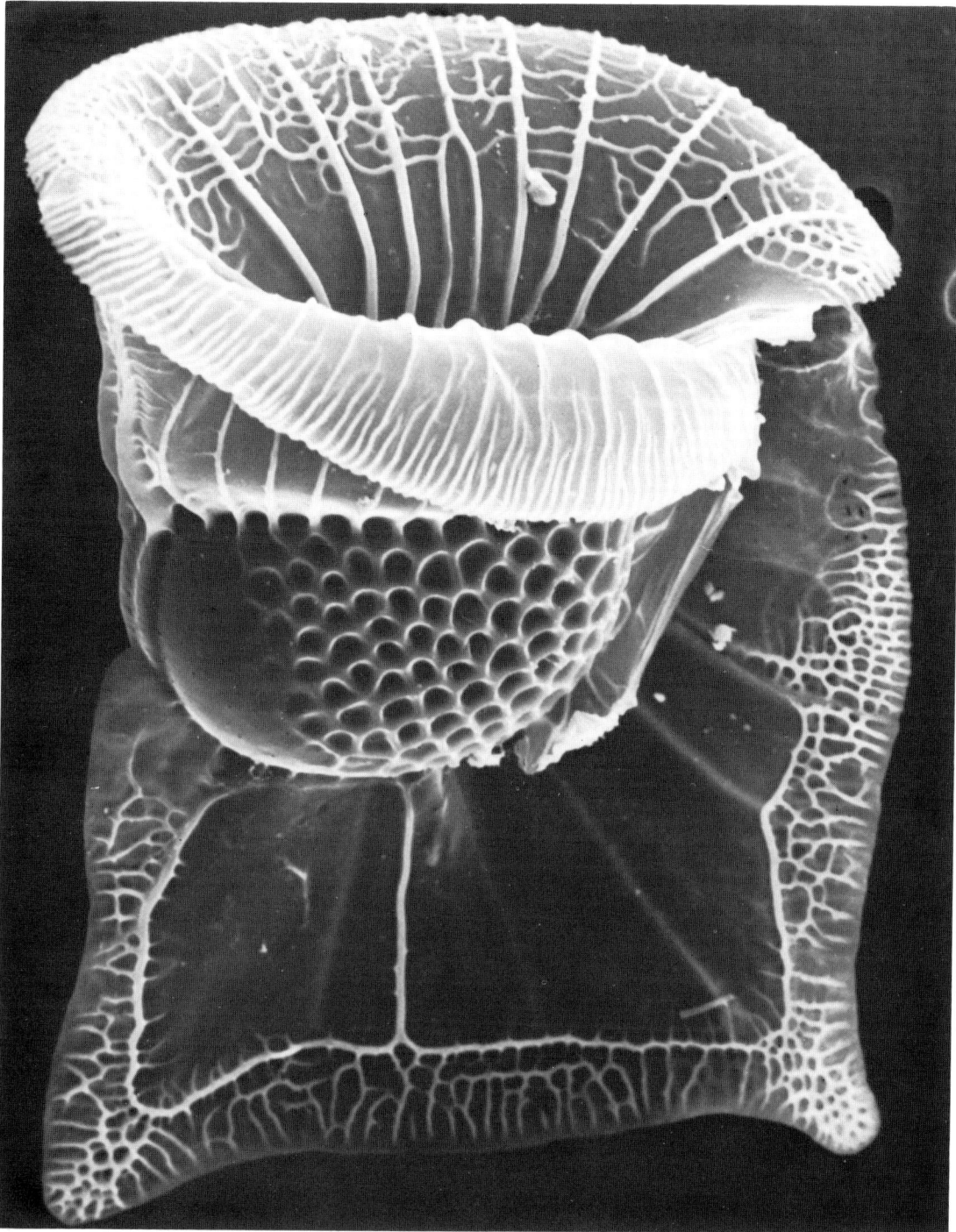

Ornithocercus quadratus Schütt
One of the many very impressive dinoflagellates which are to be found in the plankton of warm tropical seas. This genus has hugely exaggerated lists around the girdle and a very much widened sulcal list, the shape of which varies from species to species.
E. Mediterranean Sea. 130 μm l, 100 μm w.

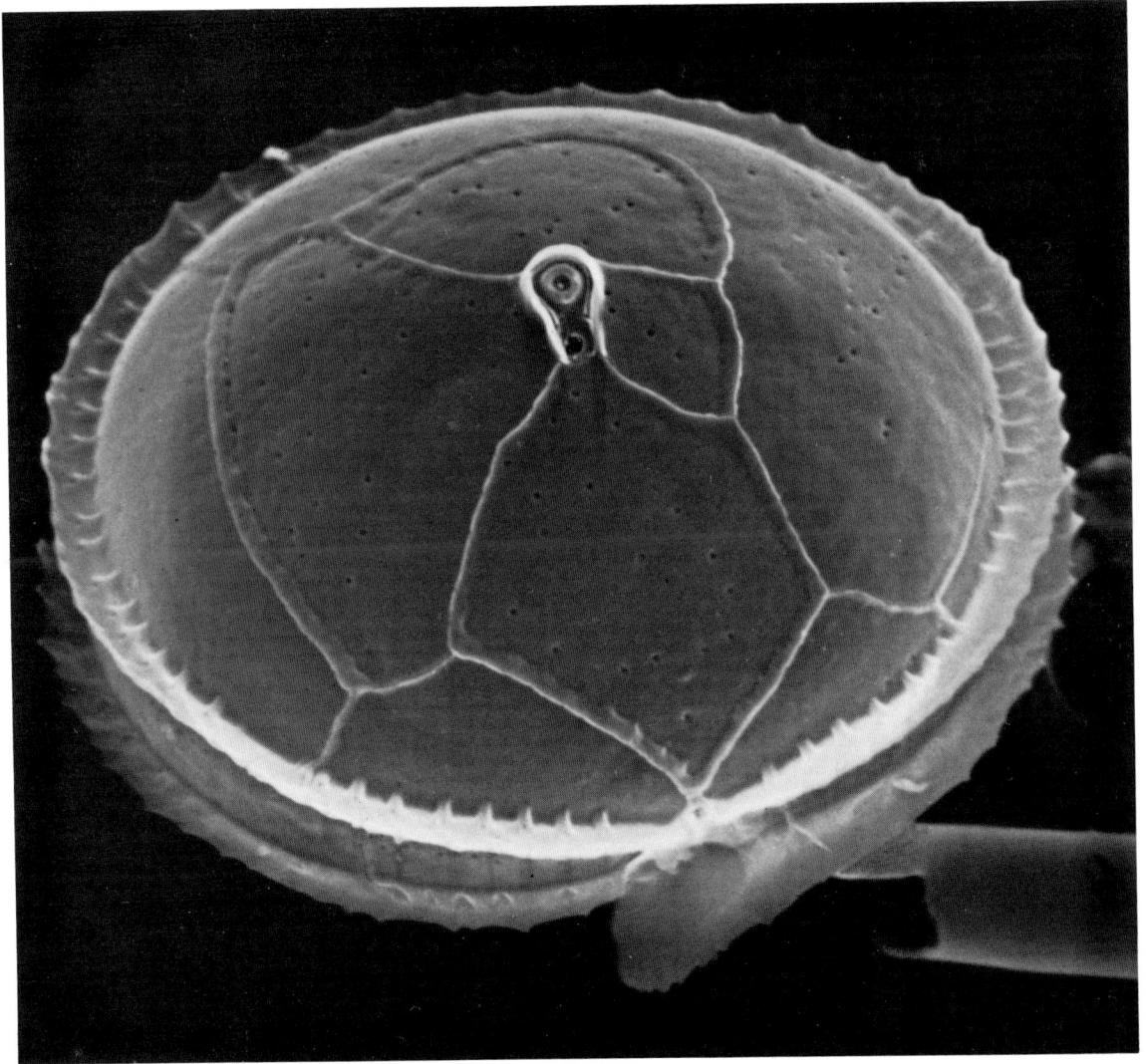

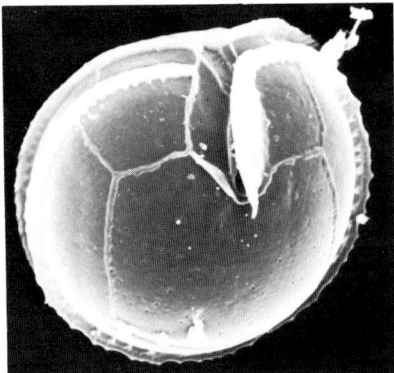

Boreadinium pisiformis Dodge &
Hermes
A small sub-spherical dinoflagellate
belonging to the Diplopsalis group. Note
the five-sided first apical plate and, in
the small picture, the single antapical
plate and sulcal wing.
Plate formula Po, 4', la, 7'', 3c, 5''', 1''''
North Sea. 35 μm l, 45 μm w.

30

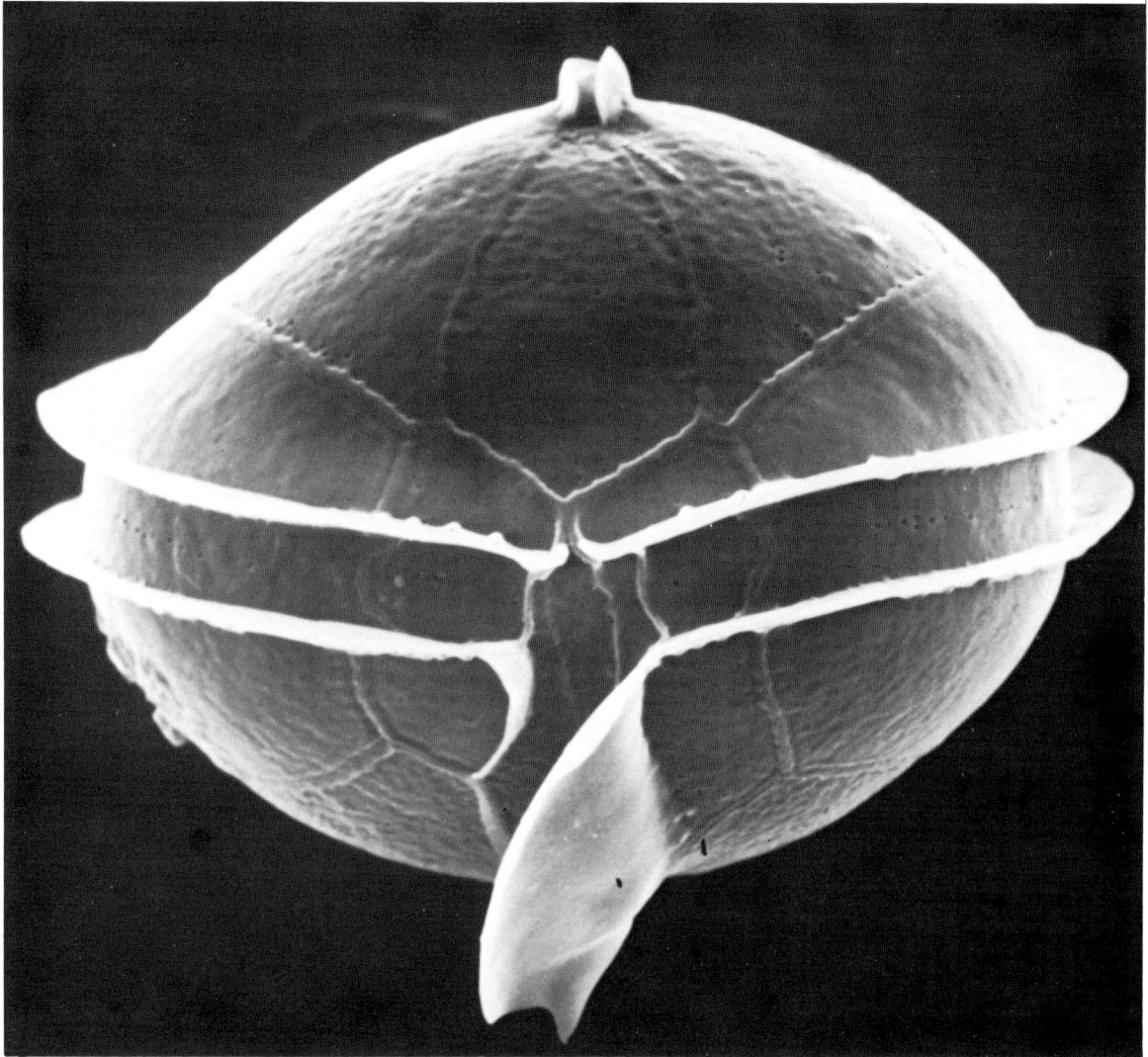

Diplopsalis lenticula Bergh
A small, non-photosynthetic dinoflagellate which is sub-spherical in shape and has a prominent sulcal wing. The four-sided first apical plate (top and small picture) distinguishes this species from some related organisms, such as *Boreadinium* and *Oblea*.
North Sea. 30 μm l, 40 μm w.

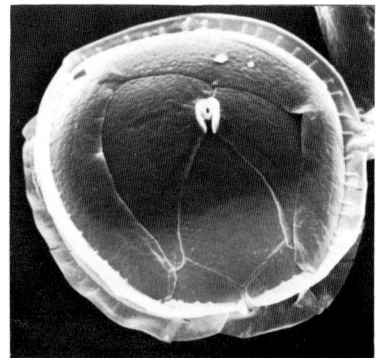

31

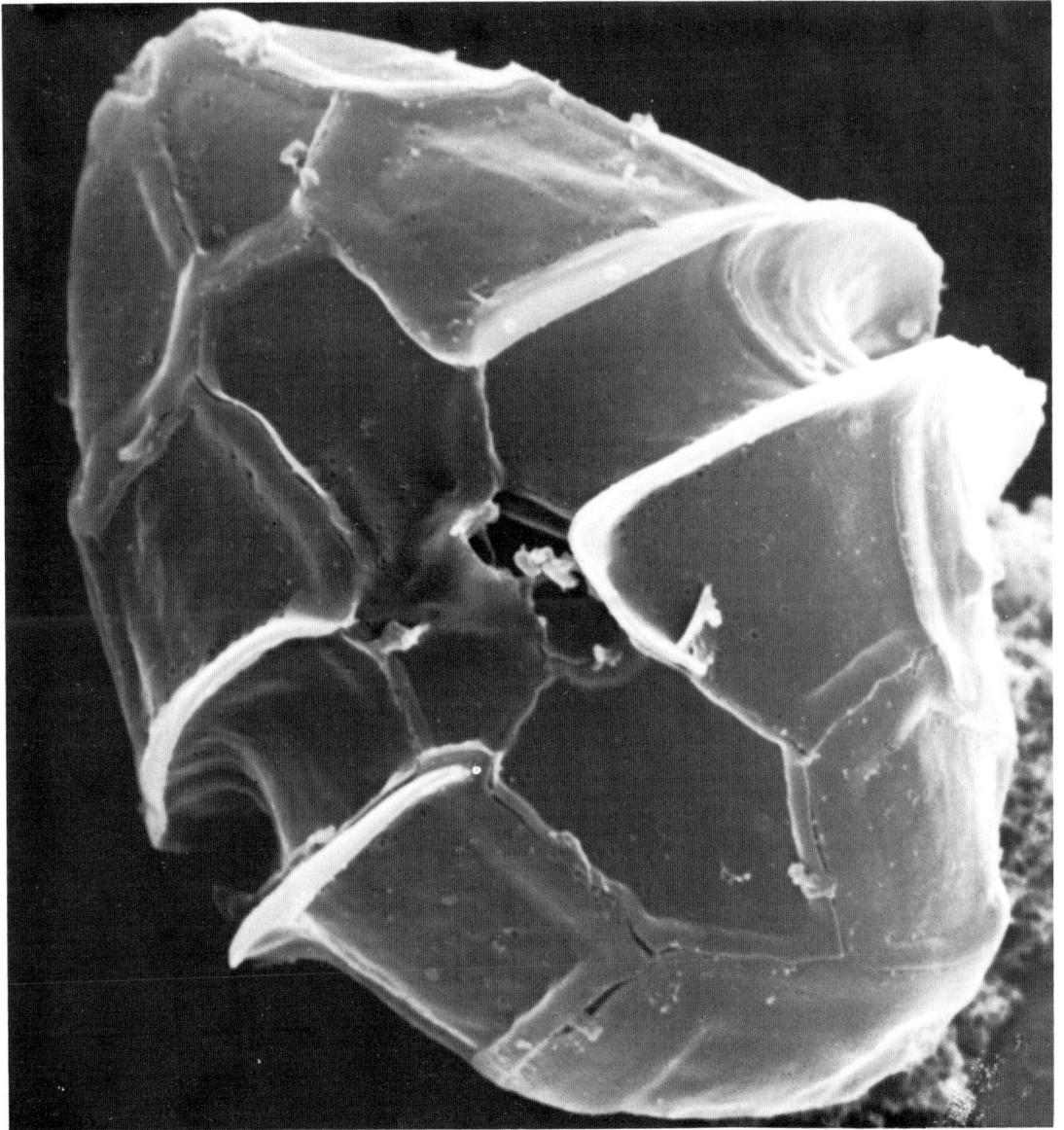

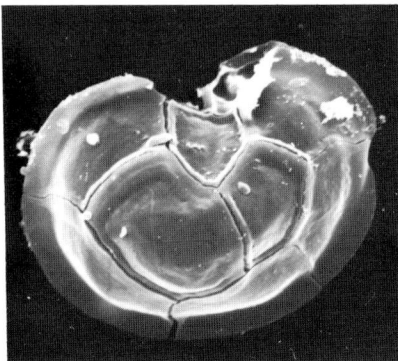

Heterocapsa triquetra (Ehrenberg)
Stein
This tiny photosynthetic dinoflagellate is
sometimes quite common in neritic situations and partly enclosed waters such as
sea lochs. As can be seen, the plates are
smooth and unornamented and the sulcal
area appears unusually simple. The
hypotheca is shown in the small picture.
North Sea. 20 μm l, 15 μm w.

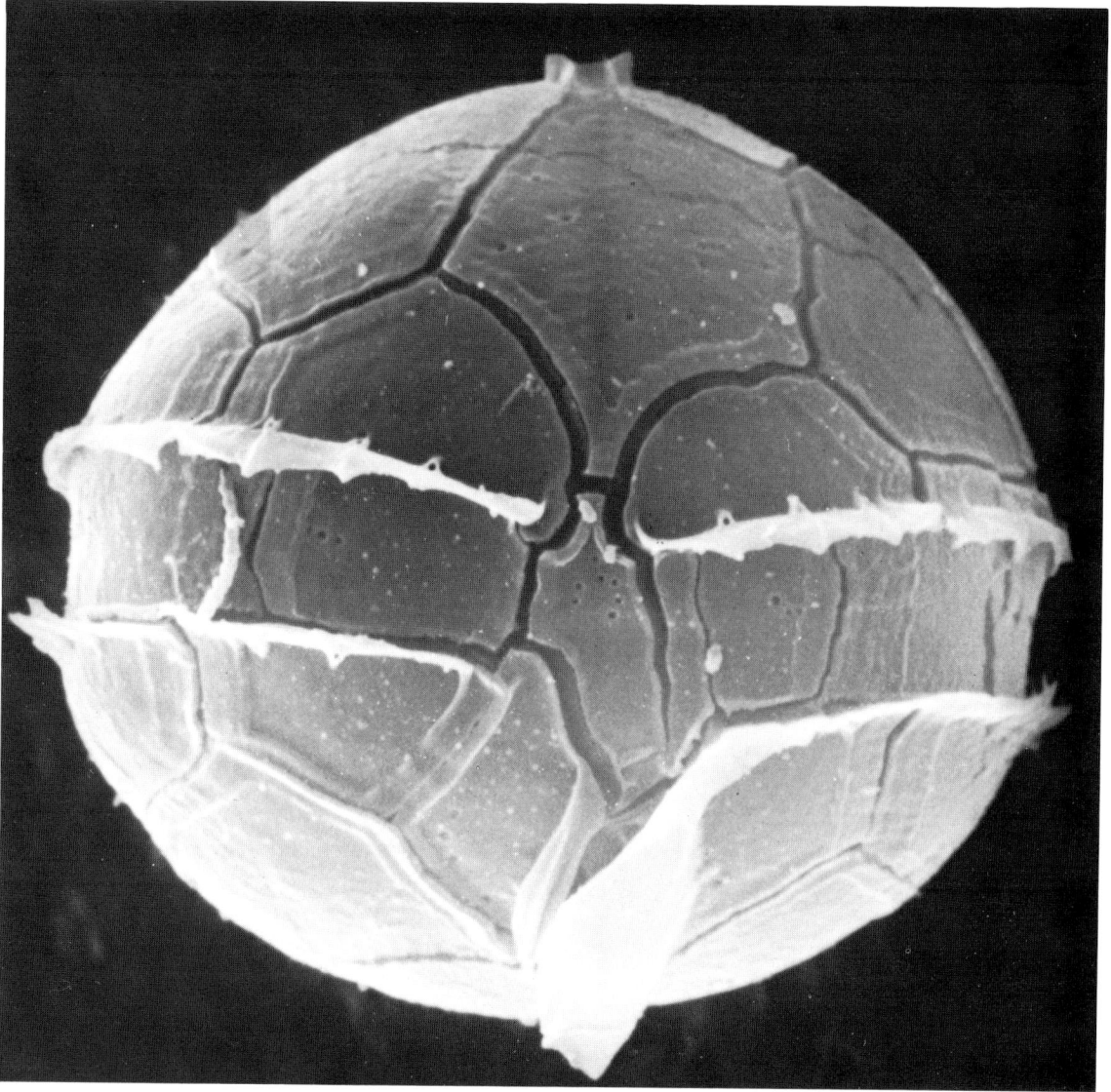

Oblea rotundata (Lebour) Balech
A very small oceanic dinoflagellate which is distinguished by its wide sulcal list, fairly wide ridged girdle lists (partly collapsed here) and two antapical plates. The first apical plate is clearly five-sided. Compare with *Boreadinium* and *Diplopsalis*.
E. Atlantic. 20 μm d.

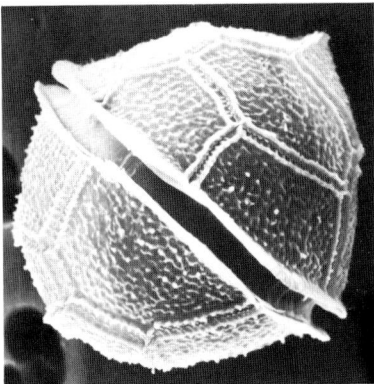

Peridiniopsis borgei Lemmermann
A freshwater dinoflagellate with thick,
ornamented, plates and a very distinctive
tabulation shown here in the ventral
(top) and dorsal (small picture) views.
Note the large flap (list) on the right side
of the sulcus.
Lake Malmö, Sweden (FW). 35 μm l,
35 μm w.

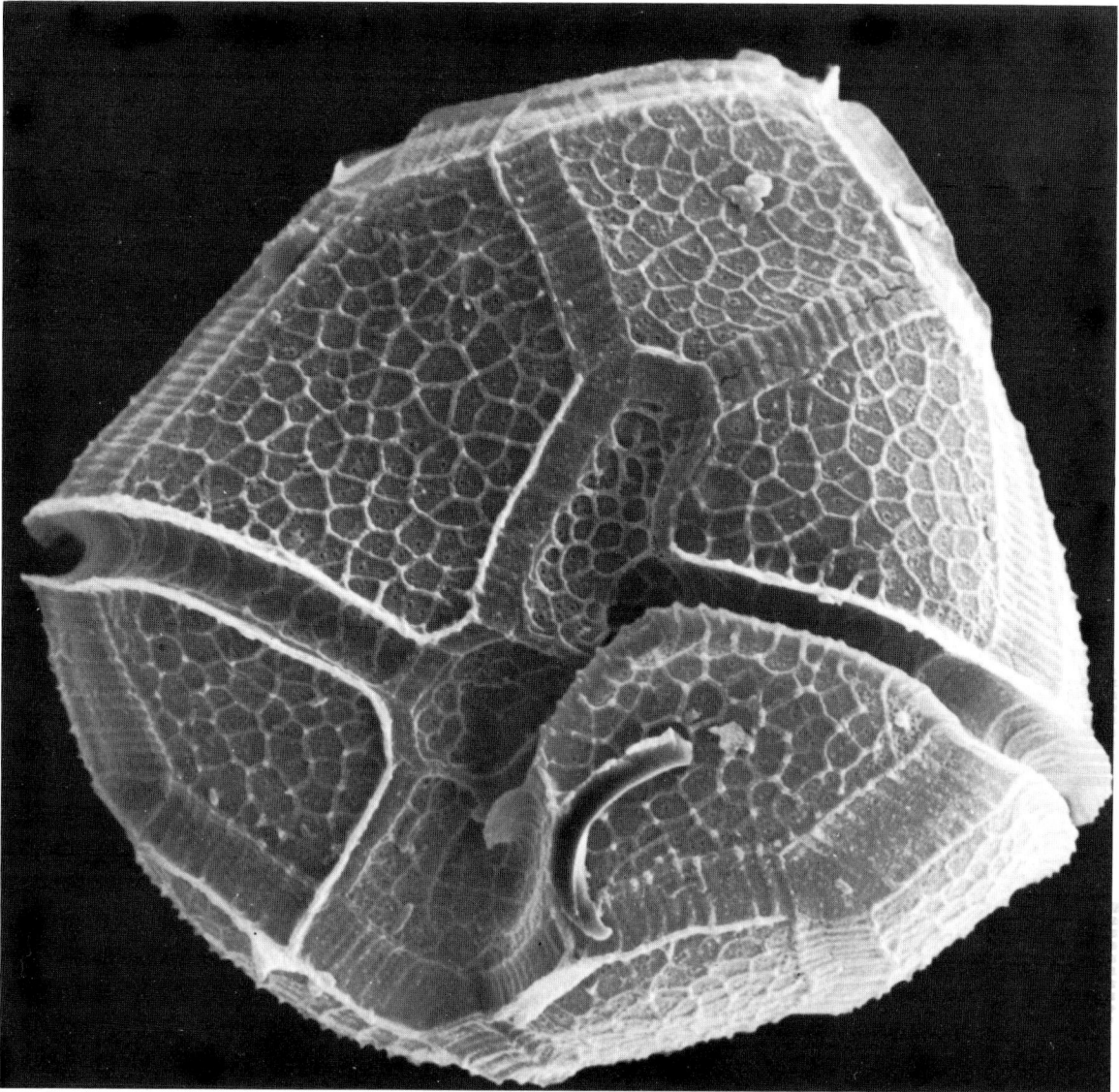

Peridinium bipes Stein
A common freshwater species found in lakes. Note the very broad first apical plate and what appears to be a rather simple sulcus of three plates. The girdle is displaced rather more than is usual in this genus. In the small picture the arrangement of the epithecal plates can be seen.
Marburg, Germany (FW). 50 μm l, 50 μm w.

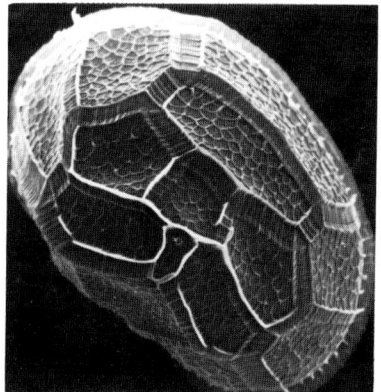

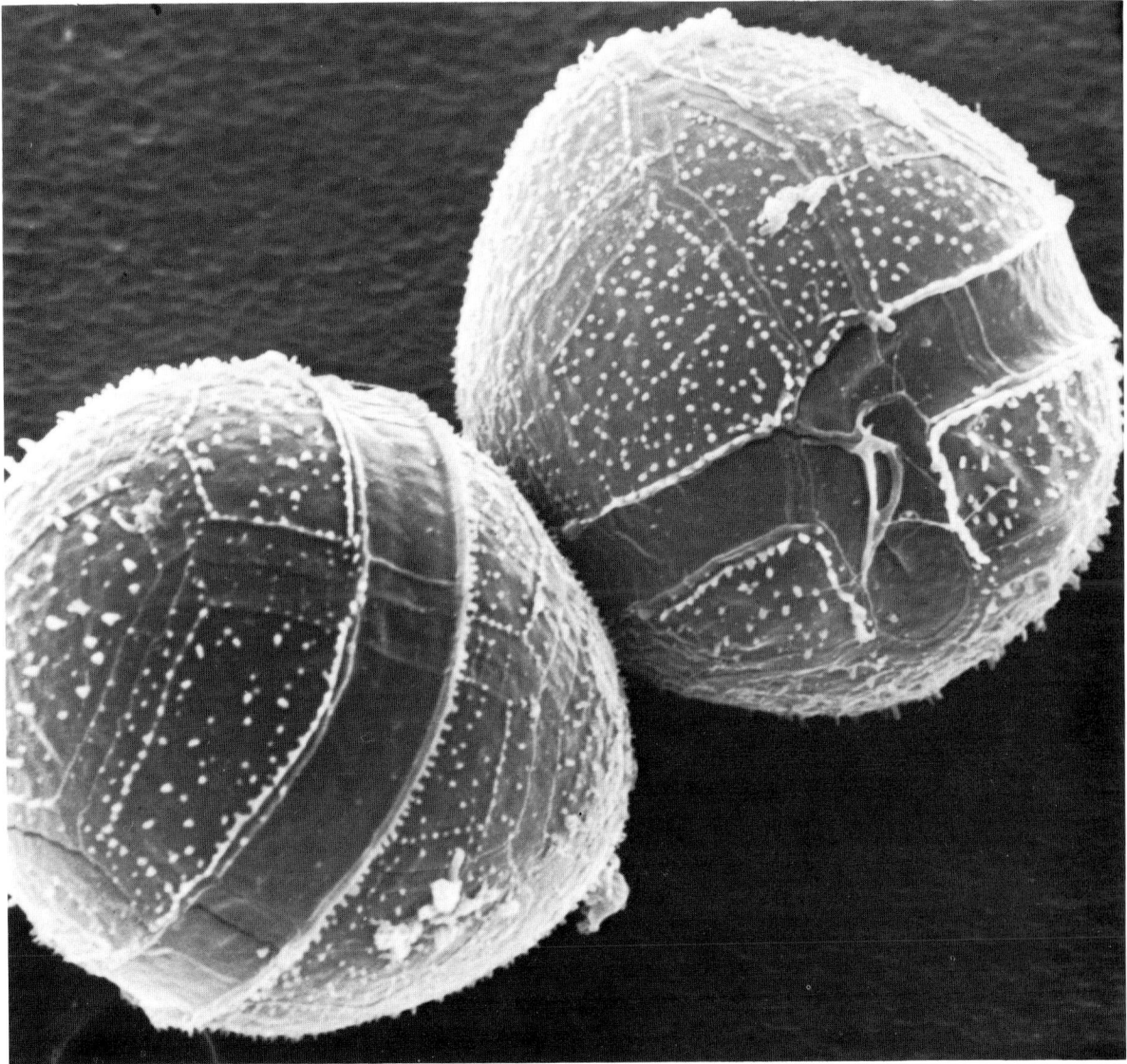

Peridinium lomnickii Lemmermann

A small species of *Peridinium* which has been found in large numbers in small bodies of water. Note the contrast between the smooth girdle and sulcus plates and the pimpled epi- and hypothecal plates.

Tarn in Cumbria (FW). 25 μm l, 22 μm¯ w.

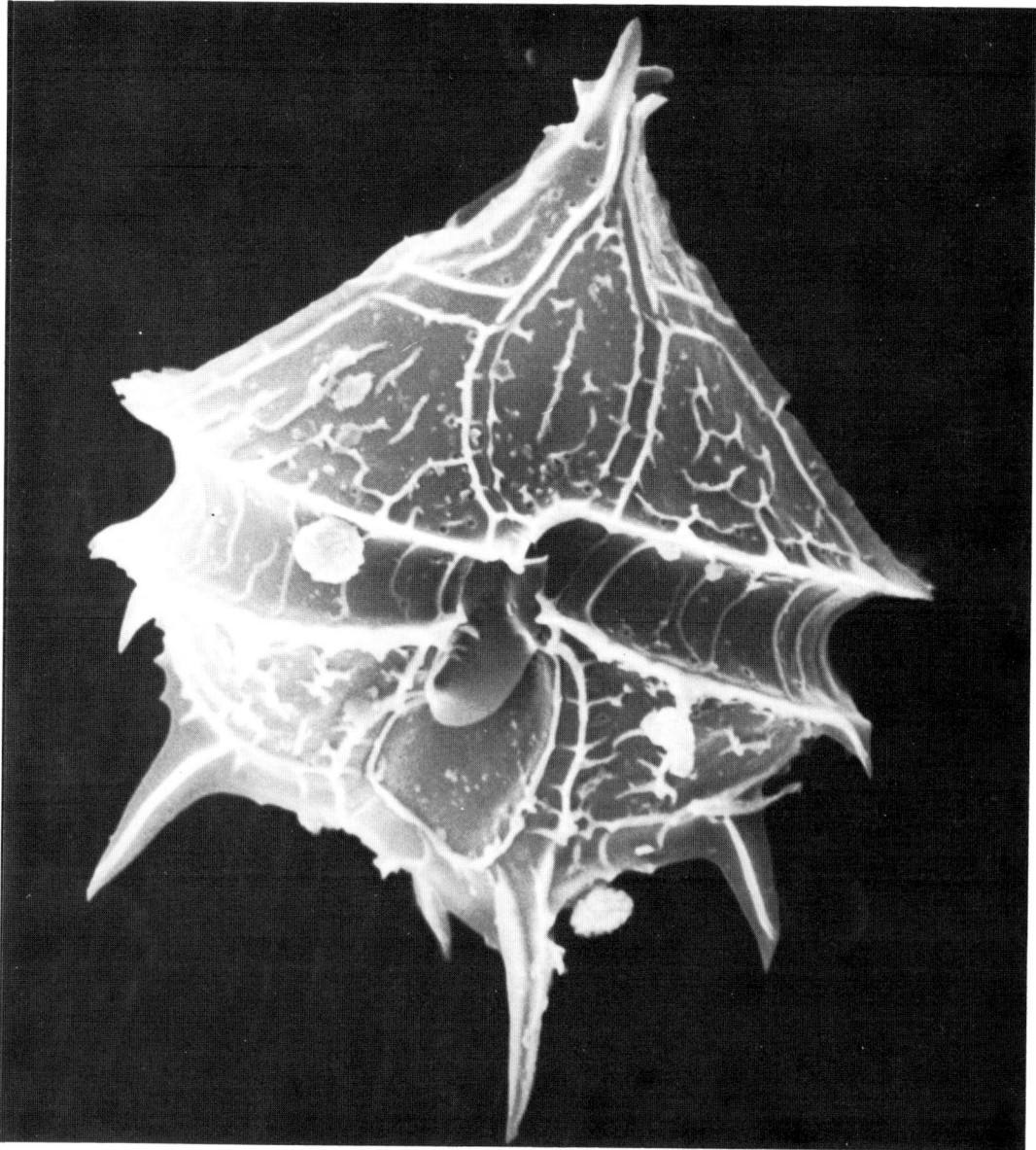

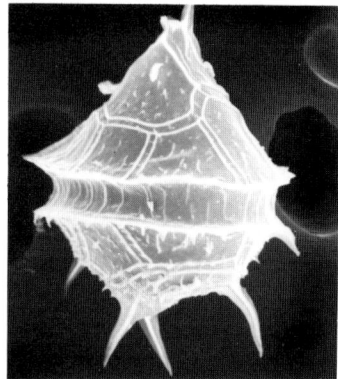

Peridinium quinquecorne Abé
An unusually spiny species which is
sometimes confused with *Gonyaulax triacantha*. The plate pattern is clearly quite
distinct and also differs from *Protoperidinium*. The small picture shows a dorsal
view.
Plate formula: Po, 3', 2a, 7'', 5c, s, 5''',
2''''.
Brackish water, S. India. 26 µm l, 20 µm w.

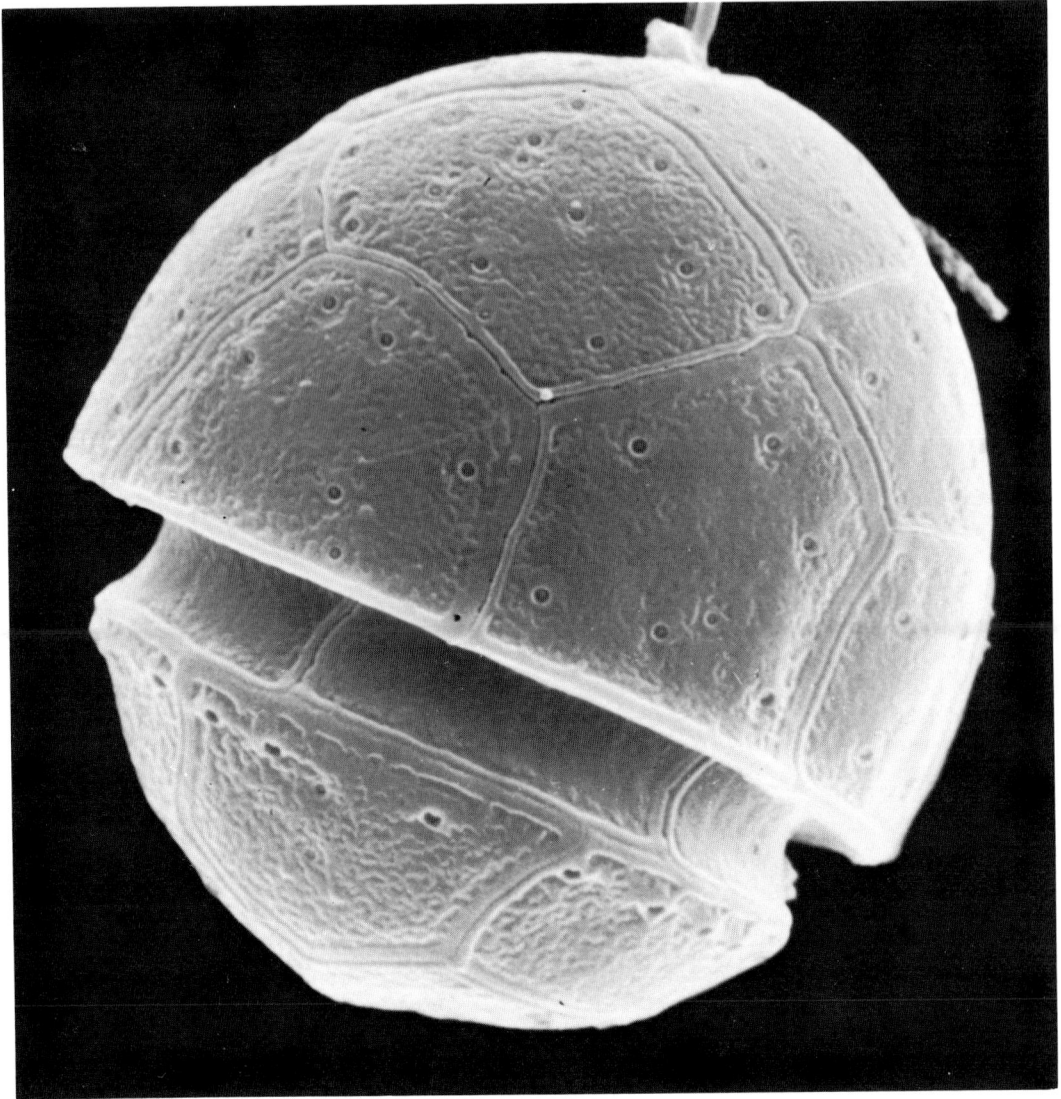

Peridinium umbonatum Stein
Dorsal view of a rather small and unusual *Peridinium* in which the hypotheca is much reduced in size. Compared to many other species of this genus the plates are very smooth. Small pond in N. Scotland (FW). 15 μm l, 13 μm w.

Protoperidinium Bergh

A set of micrographs to illustrate the four main views of the *Protoperidinium* theca. **A.** The front or ventral view in which the four-sided first apical plate is clearly visible (*P. matzenaueri*). **B.** the back or dorsal view (*P. leonis*). **C.** The top or apical view showing the apical pore and all 14 plates of the epitheca (*P. leonis*). **D.** The bottom or antapical view showing the two large antapical plates and the five smaller postcingulars (*P. ovatum*). The characteristic tabulation in this genus is: Po, 4', 3a, 7'', 3c, 5s, 5''', 2''''.

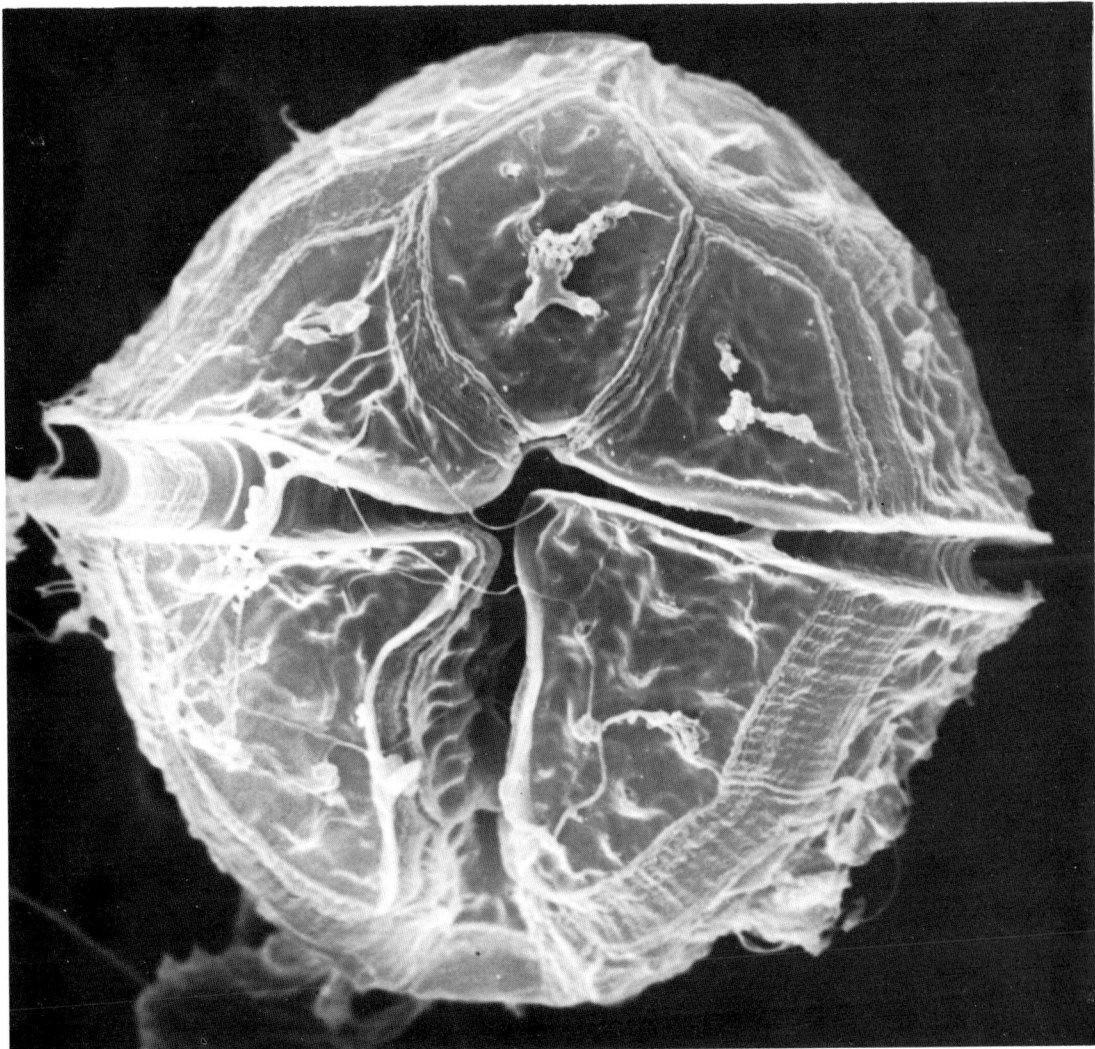

Protoperidinium avellana (Meunier)
Balech
A somewhat ovoid species with a median
girdle. The thecal plates have a rather
undulating surface and the girdle is
ridged. Note the shield-shaped first apical plate. The cyst (small picture) is
round and brown and has a distinctive
archeopyle (excystment aperture).
W. Scotland. 47 µm l, 50 µm w.

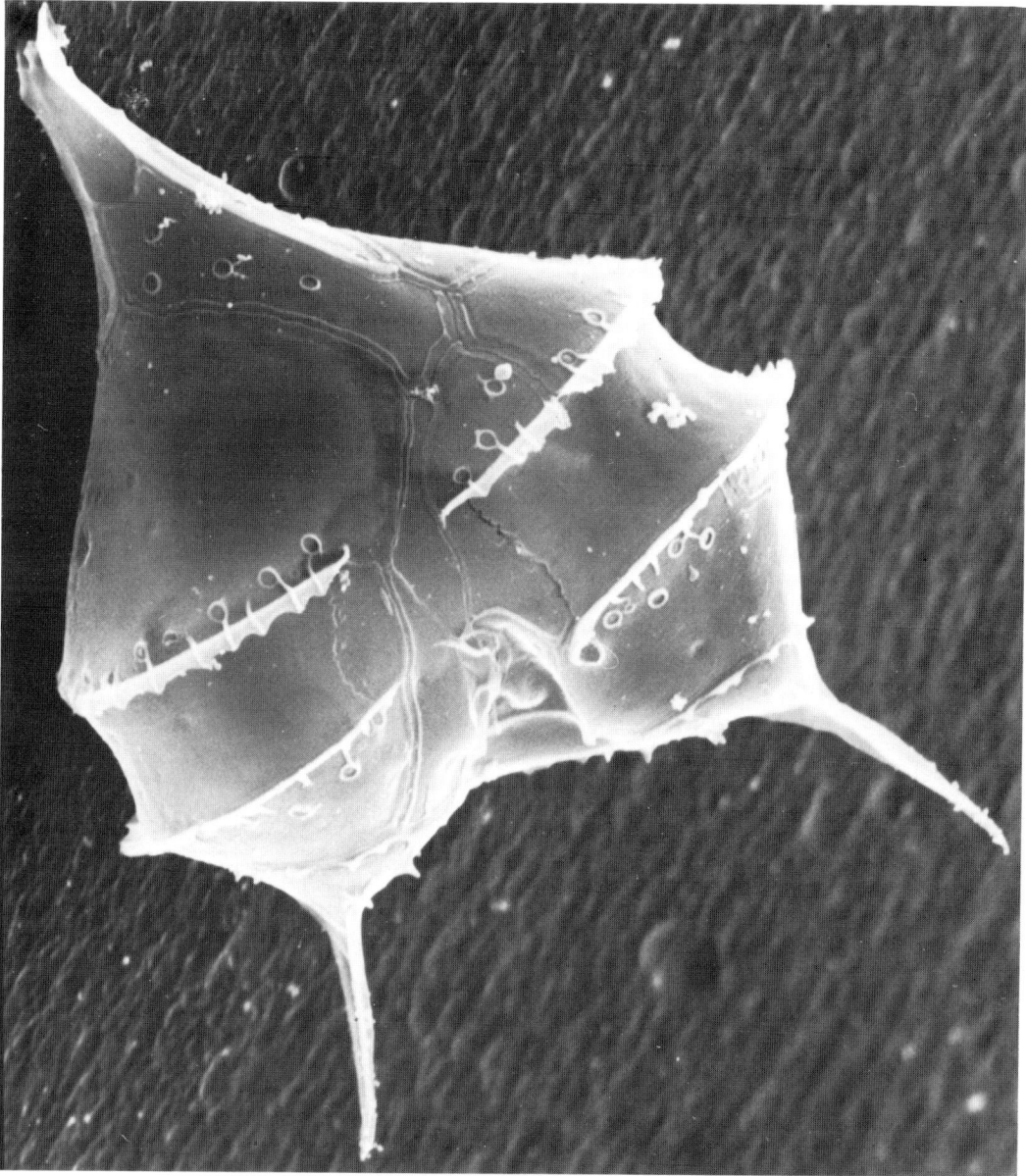

Protoperidium bipes (Paulsen) Balech
(= *Minisculum bipes; Peridinium minisculum*)
A delicate species of neritic waters. It has a chequered taxonomic history as a result of
having a rather unusual arrangement of the thecal plates.
Atlantic, W. of Ireland. 35 μm l, 22 μm w.

41

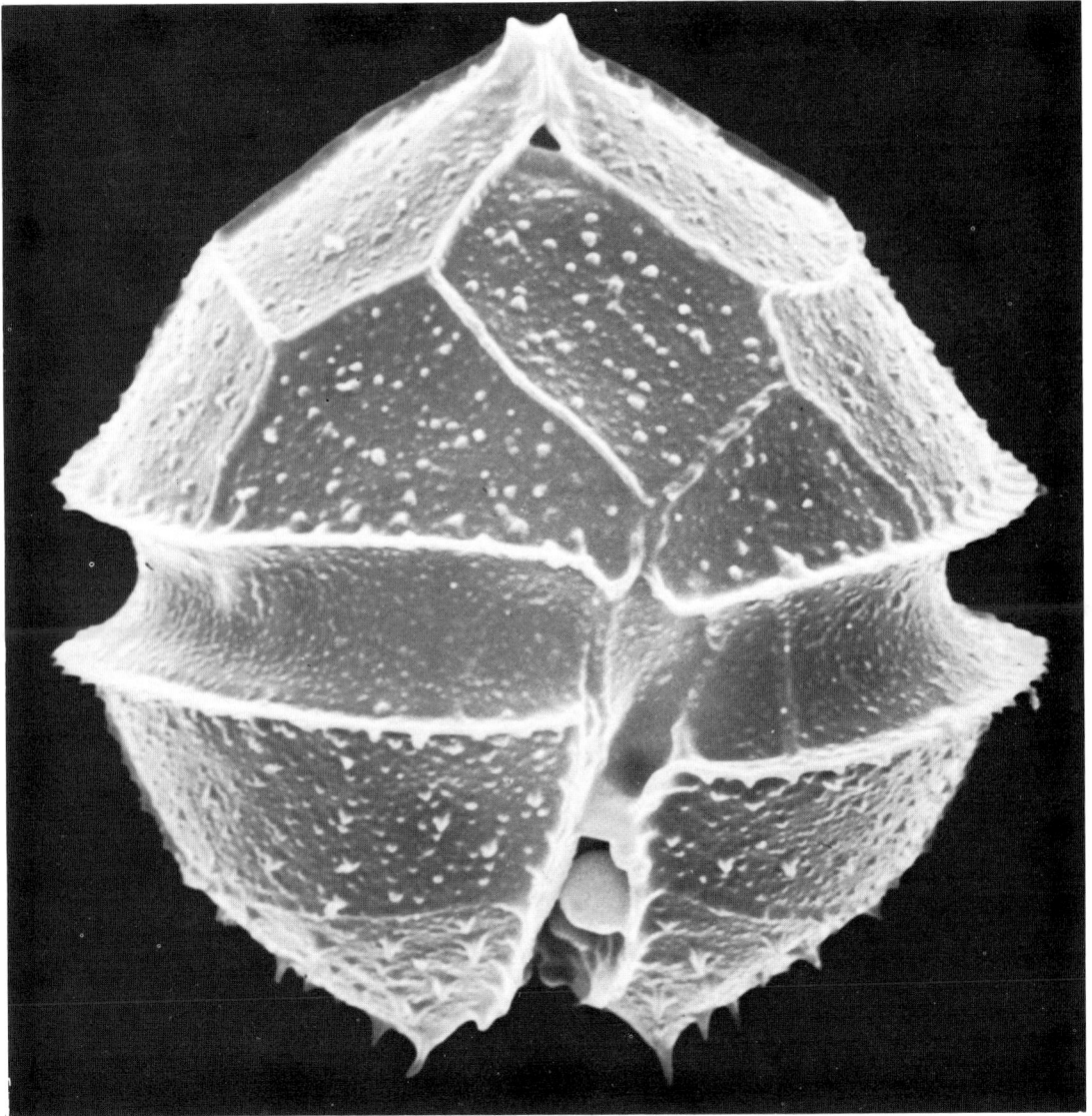

Protoperidinium brevipes (Paulsen) Balech
A fairly common though easily overlooked species with characteristic form. Note the five-sided (meta) first apical plate and the very simple ornamentation.
E. Atlantic. 24 μm l, 20 μm w.

42

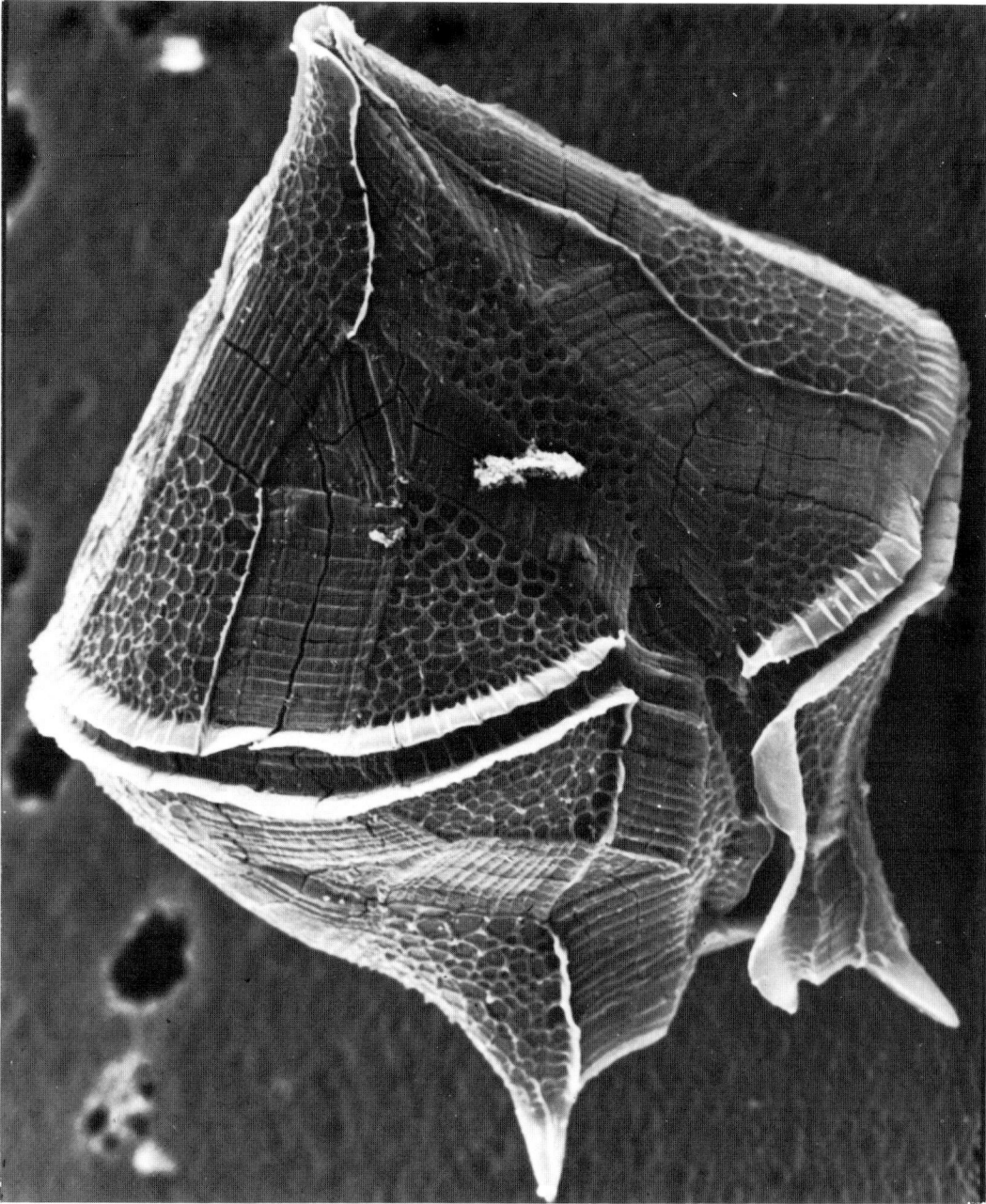

Protoperidinium brochii (Kofoid & Swezy) Balech
A large meta-*Protoperidinium* of warm waters. In this specimen the sutural bands are very
well developed and the sulcal list can be clearly seen.
Tropical E. Atlantic. 110 μm l, 80 μm w.

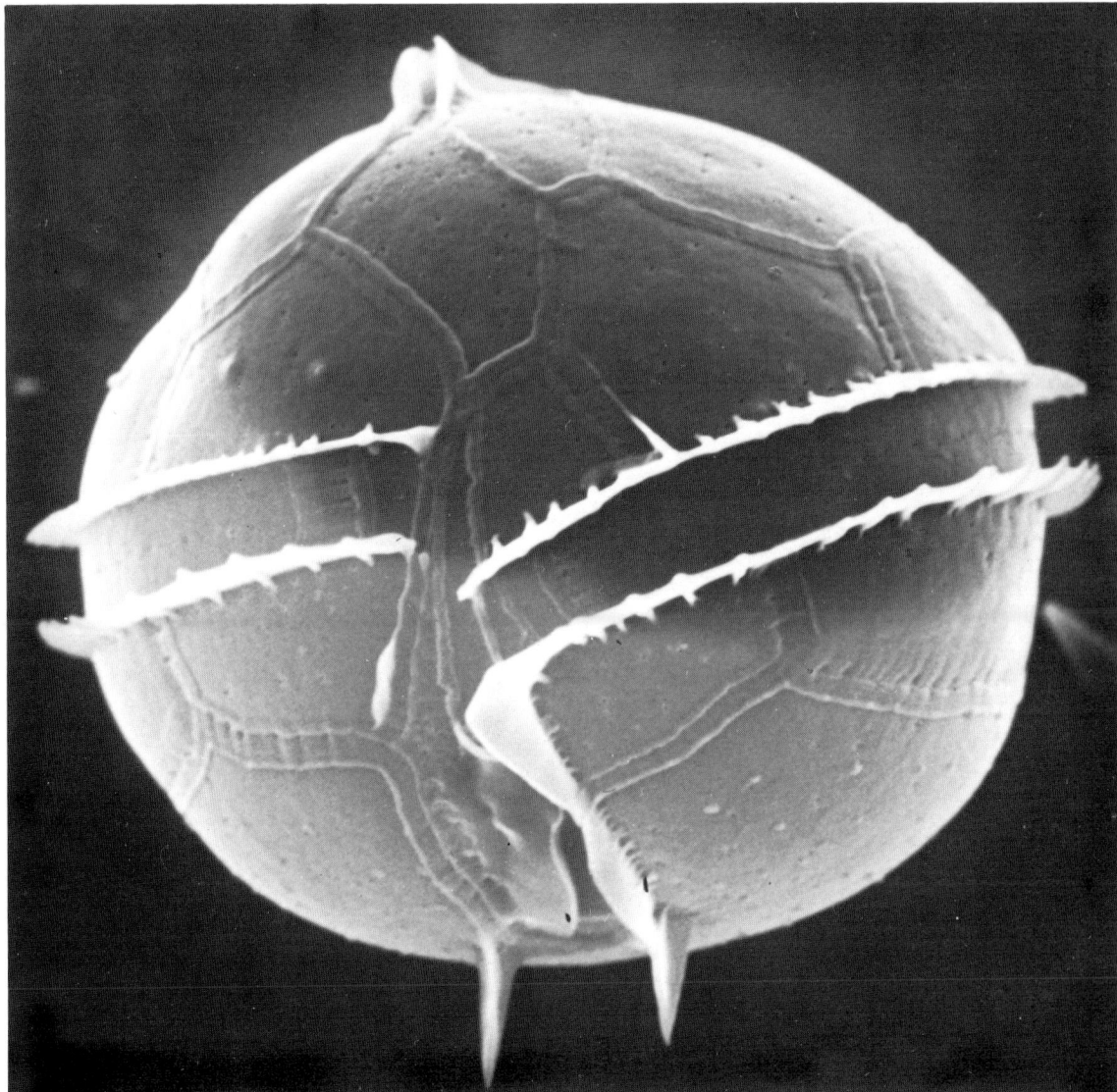

Protoperidinium cerasus (Paulsen) Balech
A fairly common meta-*Protoperidinium* species of temperate waters. The plates are virtually smooth and the girdle is displaced on the right side (contrast with *Gonyaulax*). It is distinguished by its almost spherical shape and two solid antapical spines.
North Sea. 40 μm d.

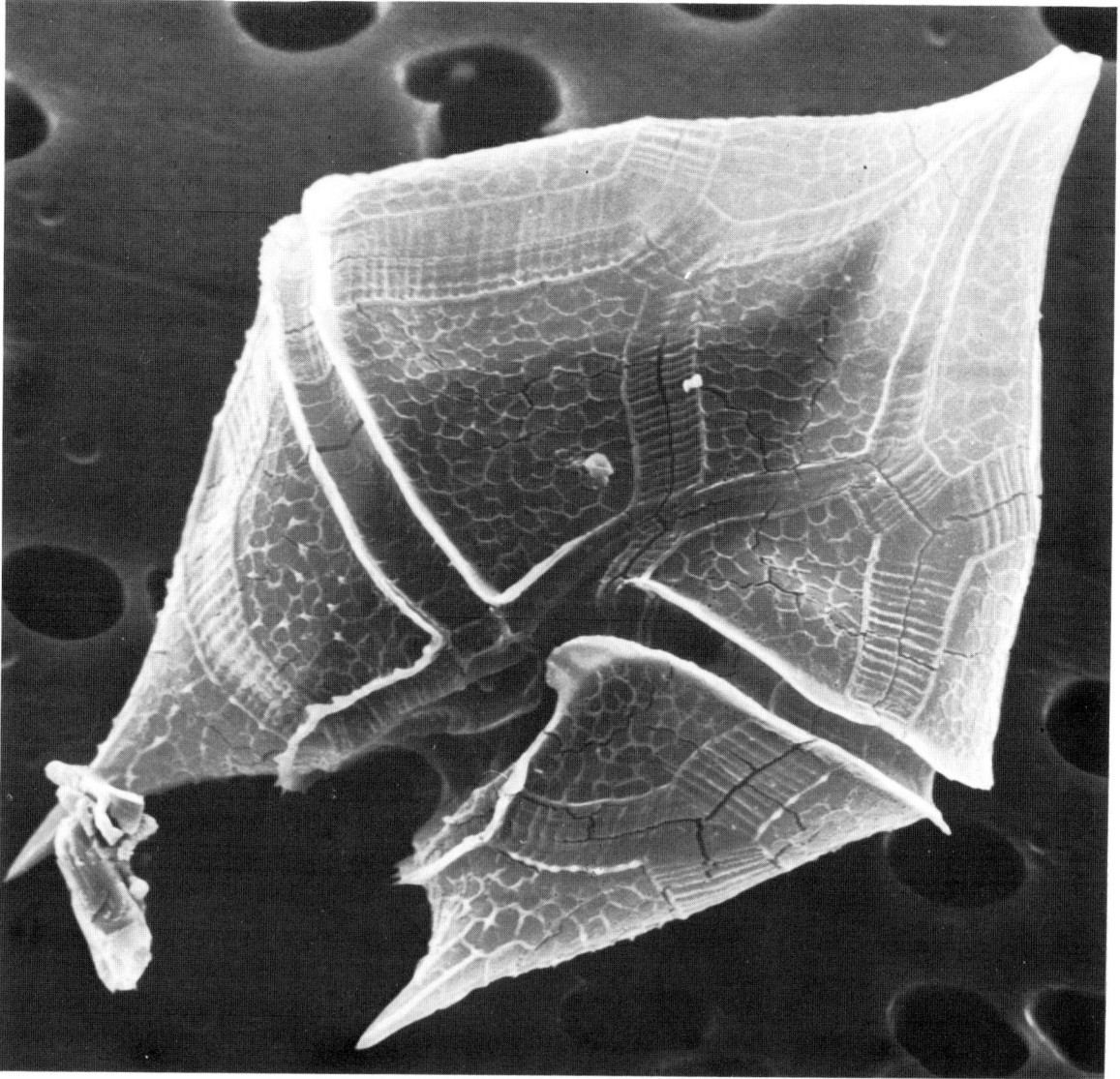

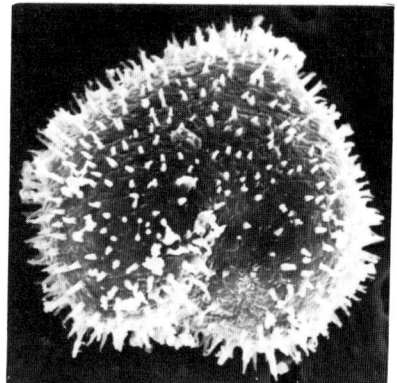

Protoperidium claudicans (Paulsen)
Balech
A fairly common planktonic dinoflagellate of temperate waters. It is similar in
its plate pattern to *P. depressum* but is
much slimmer around the girdle. There
is a spiny cyst (small picture) with a
peridinioid shape.
North Sea. 100 μm l, 70 μm w.

45

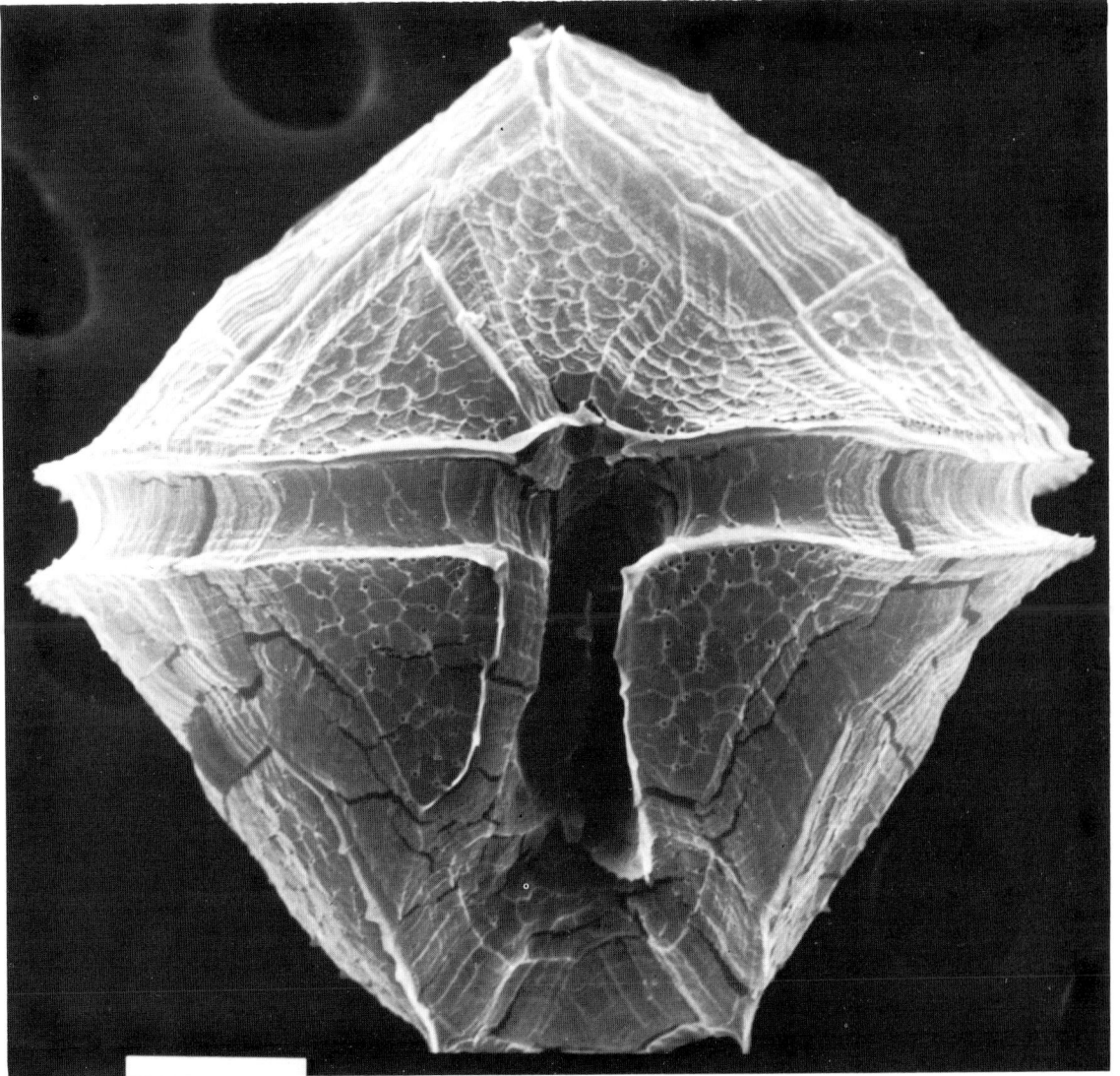

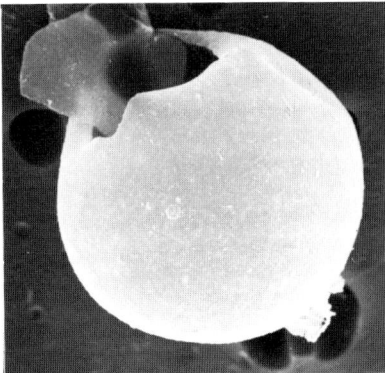

Protoperidinium conicoides (Paulsen)
Balech
A squarish species which is about as
broad as it is long. Note that the first
apical plate is four-sided and the sulcus
is fairly deeply incised. It has a round
brown cyst (small picture) with a four-
sided archeopyle.
North Sea. 56 µm l, 57 µm w.

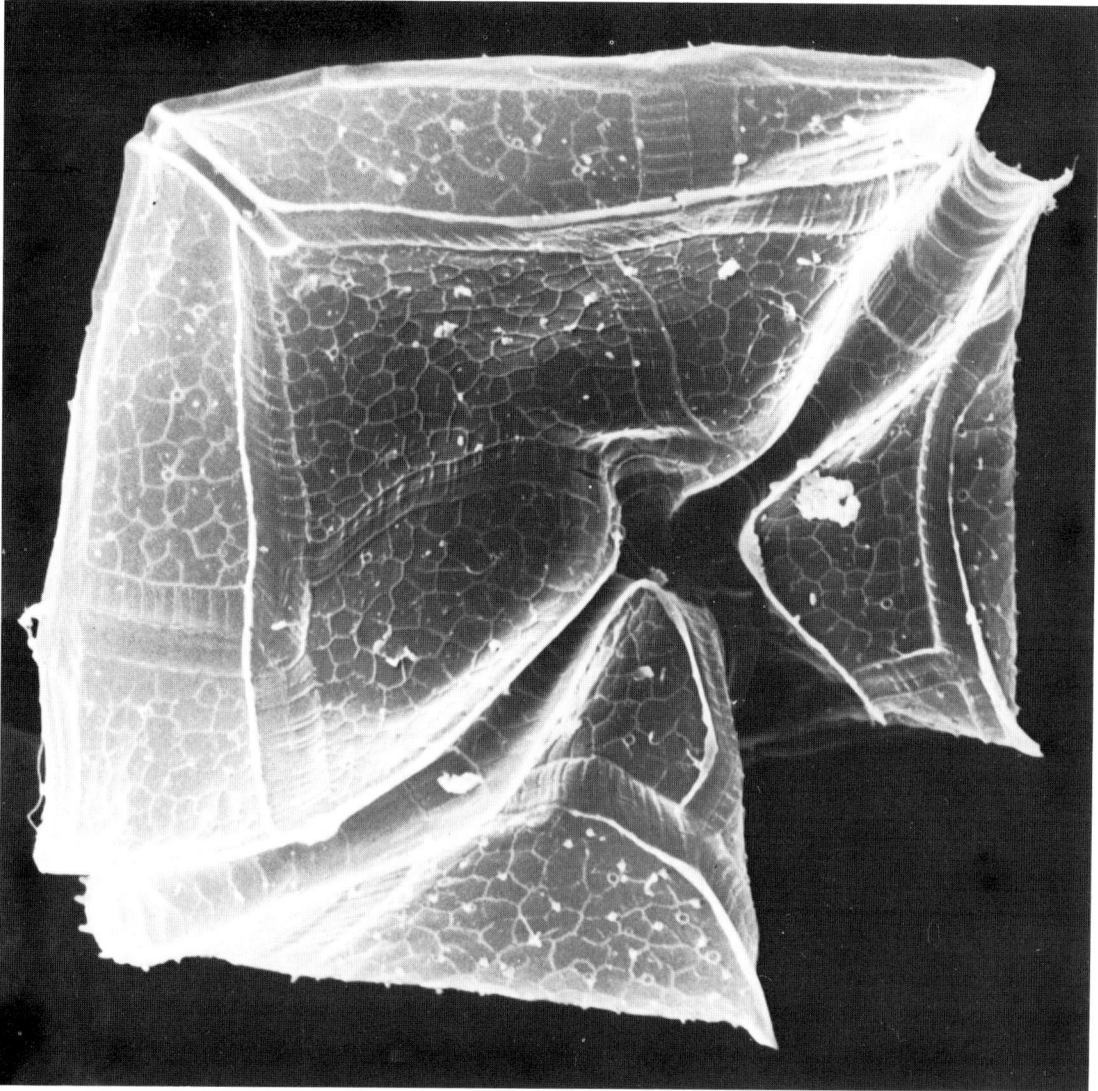

Protoperidinium conicum (Gran)
Balech
A large and distinctively shaped dinoflagellate. It has both a conical epitheca and
a conical sulcal depression in the hypotheca. The straight lines formed by the
epithecal sutures between the apical pore
and girdle also form a distinguishing feature. The spiny cyst, known as *Seleno-
phemphix* is shown in the small picture.
North Sea. 52 μm l, 58 mm w.

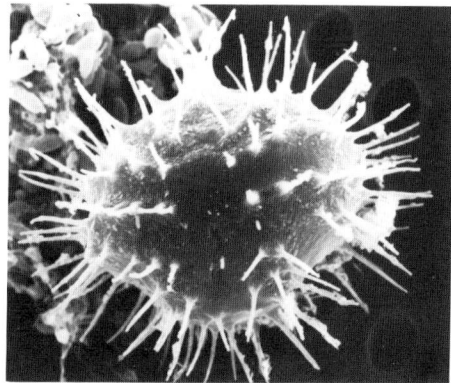

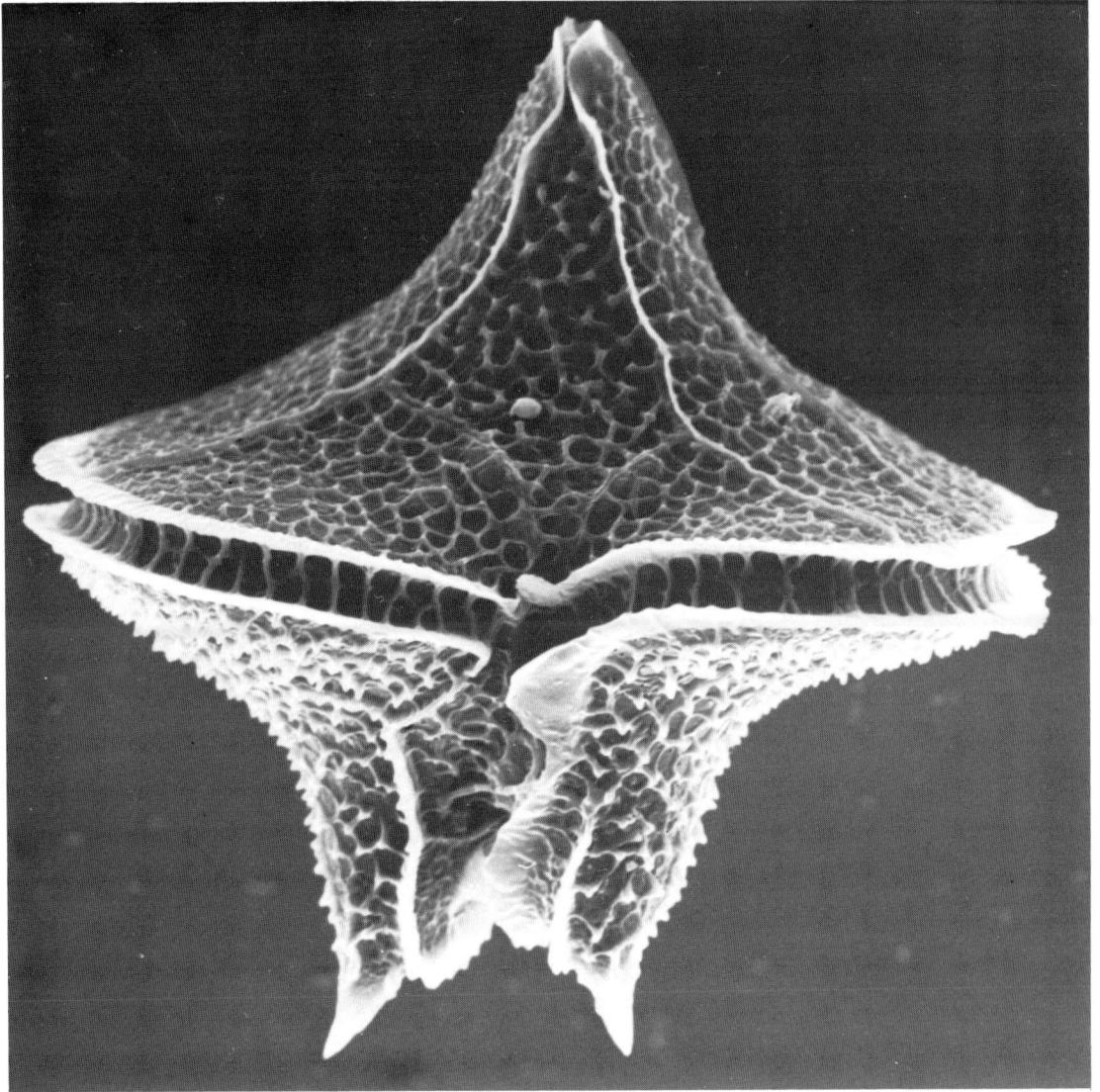

Protoperidinium crassipes (Kofoid) Balech
A fairly common species in waters to the north and east of the British Isles. It is often confused with *P. divergens* but differs in being at least as wide as it is long. Note the overall reticulate ornamentation.
North Sea. 77 µm l, 76 µm w.

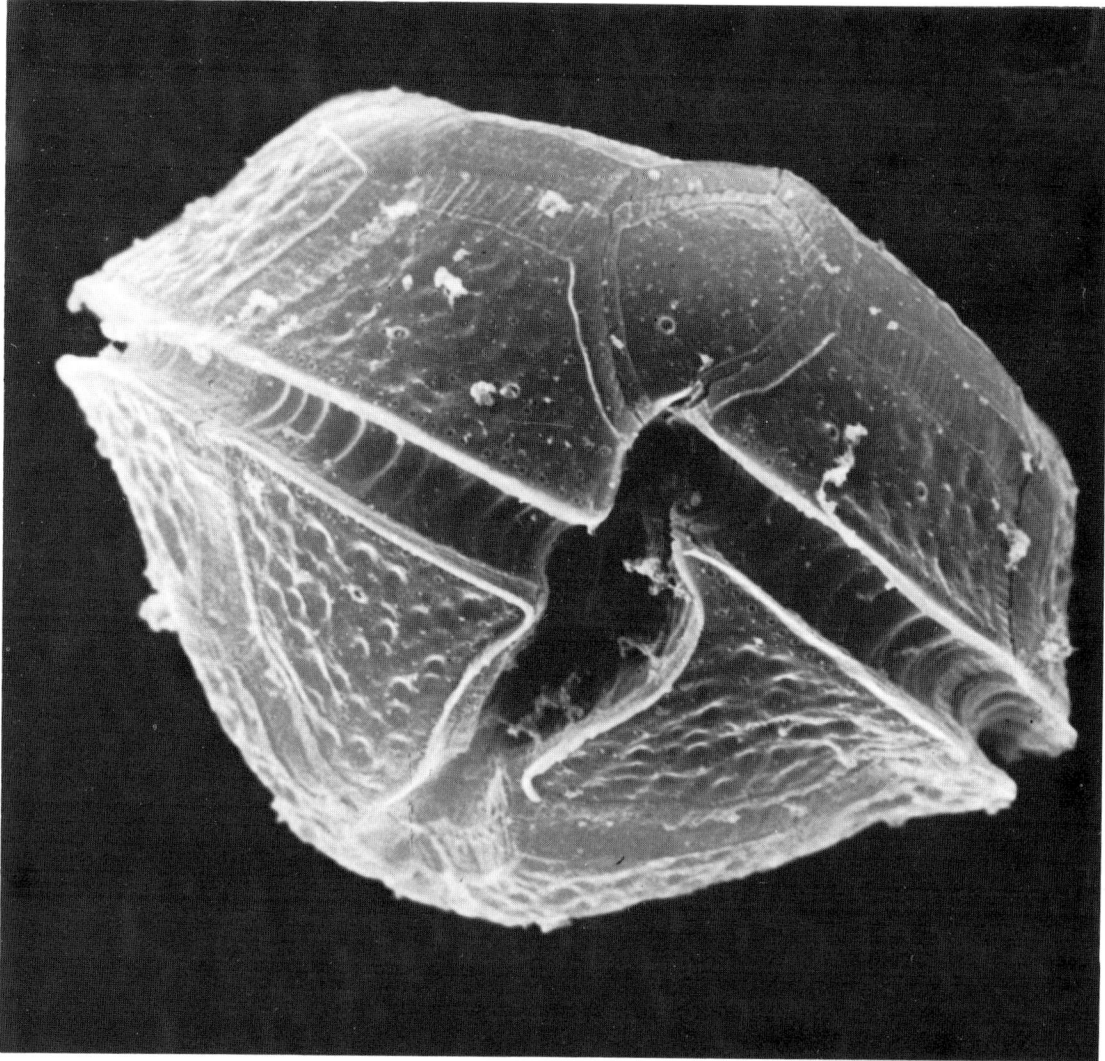

Protoperidinium denticulatum (Gran & Braarud) Balech
A very distinctively shaped species in which some cells have a flat epitheca (as here) whilst others a flat hypotheca. Often cells are found in pairs. There is a round brown cyst stage. Fairly rare, around the north of the British Isles.
North Sea. 40 μm l, 54 μm w.

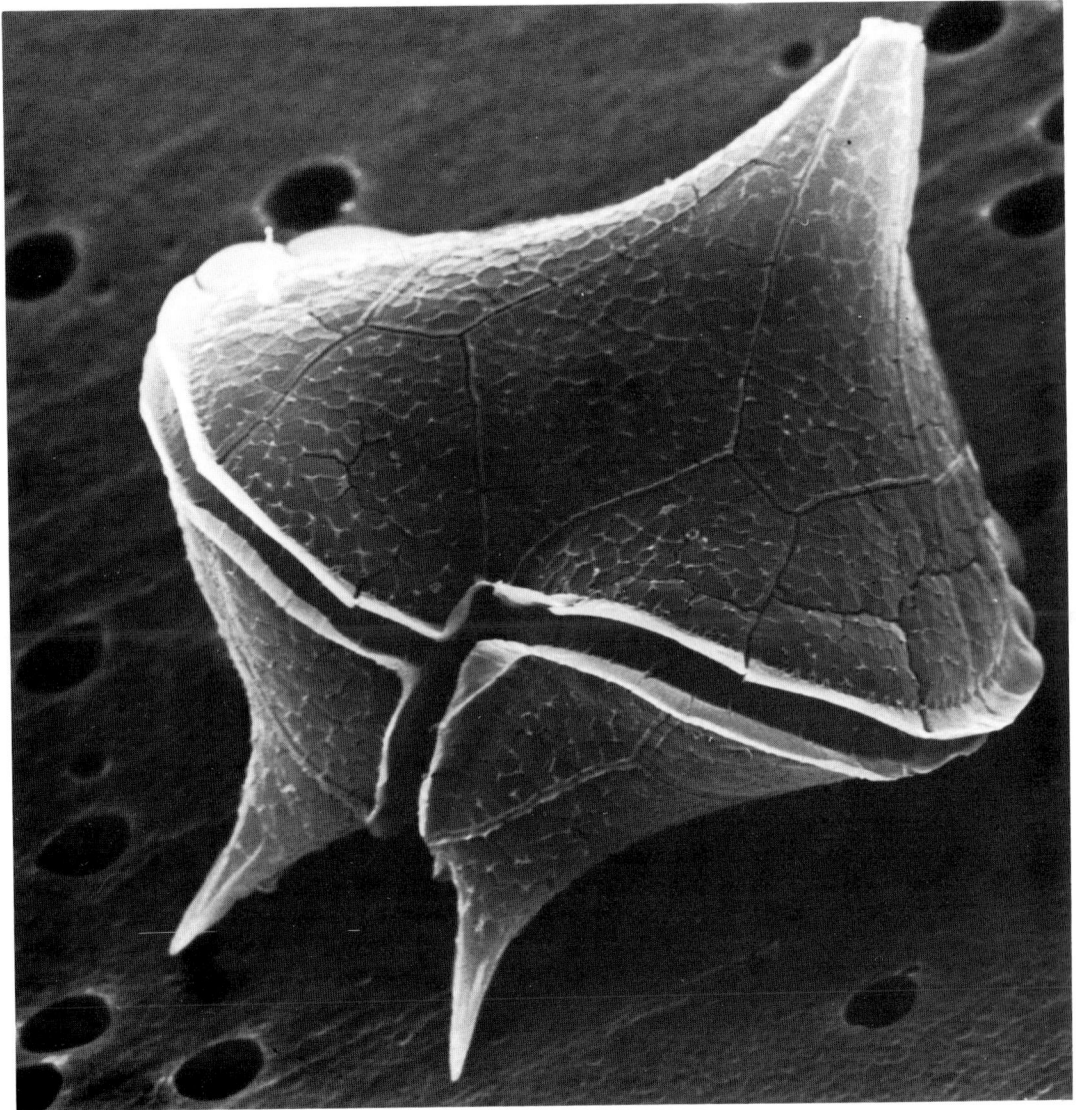

Protoperidinium depressum (Bailey) Balech
A large and distinctive dinoflagellate which is found in temperate plankton all through the year. It is distinguished from rather similar species (like *P. claudicans*) by being almost as wide, at the girdle, as the cell is long.
North Sea. 105 μm l, 85 μm w.

Protoperidinium diabolum (Cleve) Balech
A species with prominent apical and antapical horns. Note the slight displacement of the
girdle and the sulcal list which also protrudes at the posterior end of the cell.
E. Atlantic. 73 μm l, 54 μm w.

Protoperidinium divergens (Ehrenberg) Balech
Rather similar to *P. crassipes* but much slimmer at the girdle. As in *P. subinerme* the reticular ornamentation is drawn out to form tiny spines, particularly on the hypotheca. North Sea. 72 µm l, 56 µm w.

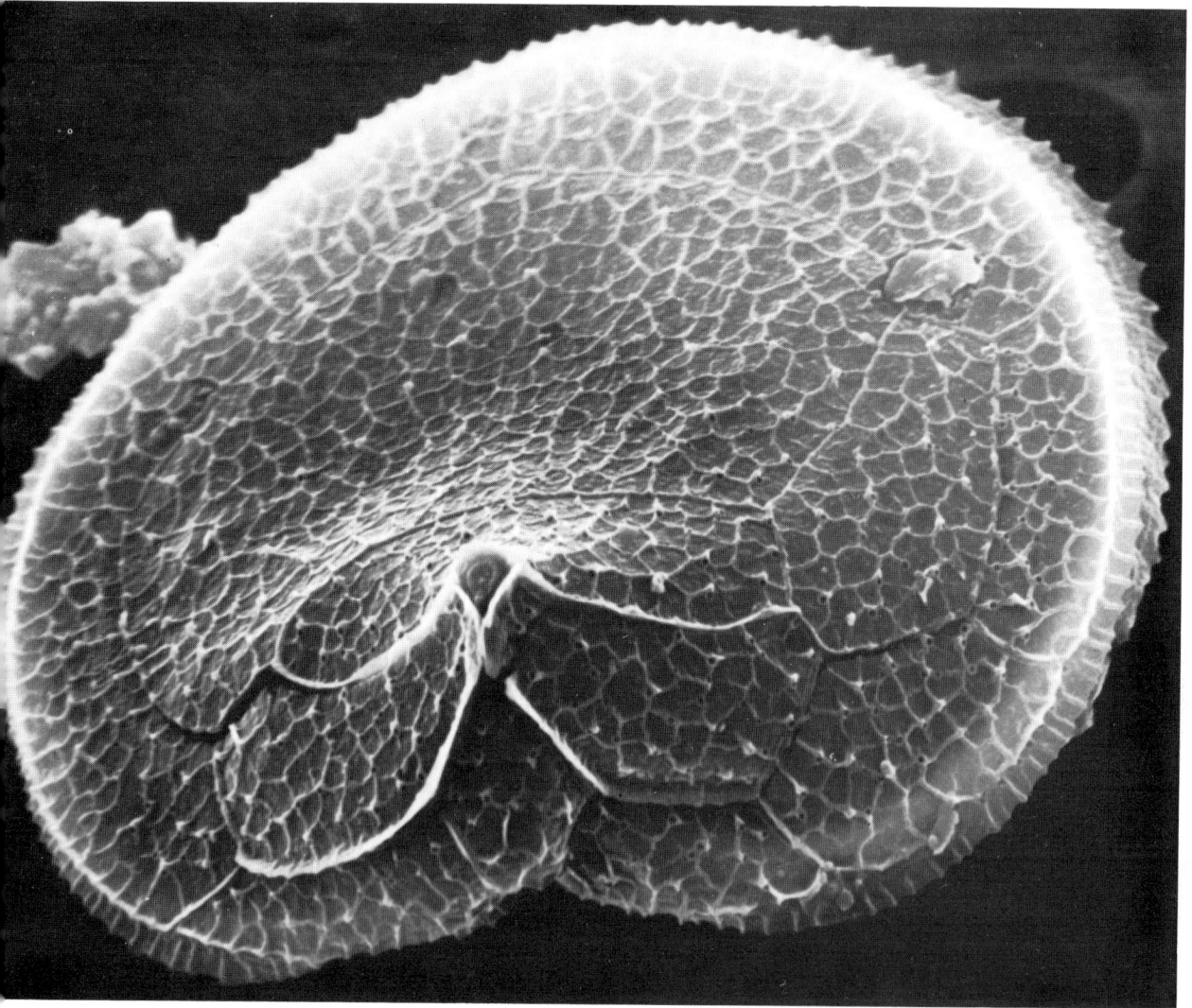

Protoperidinium excentricum (Paulsen) Balech
In this species the cell is distorted by the apex being pushed towards the ventral side (top). There are two very large intercalary plates but the apical plates are quite small. The characteristic side view is shown in the small picture.
North Sea. 35 μm l, 60 μm w.

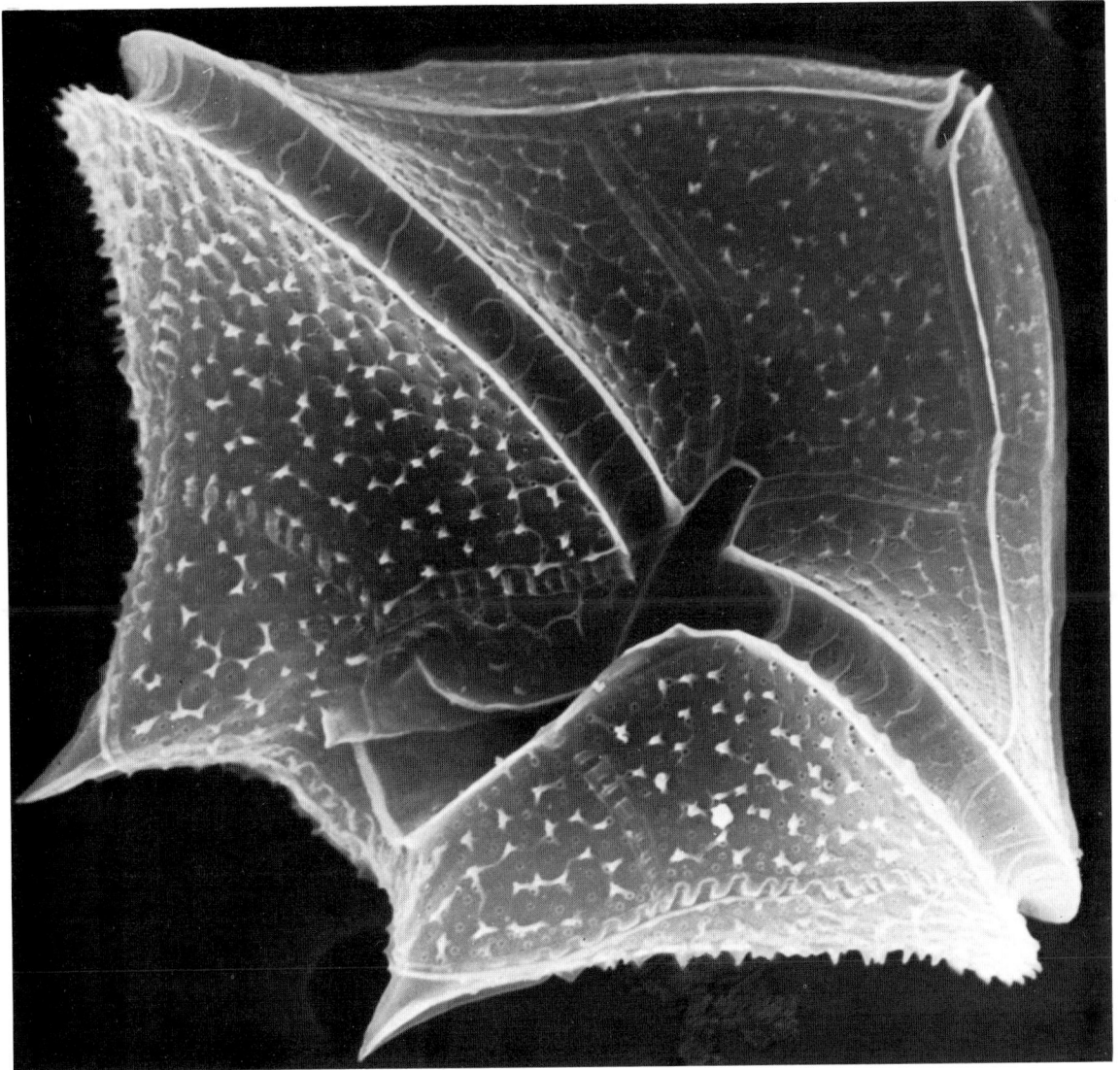

Protoperidinium exiquipes (Mangin) nov. comb.
(Basionym: *Peridinium exiquipes* Mangin 1930 *Arch. Mus. Nat. Hist. Paris*, 34 p. 377
Fig. 3)
This species has a very broad first apical plate and like *P. conicum* has straight sutures
between the apical pore and the girdle. It might be a form of *P. pentagonum* in which the
centre of the ventral side of the cell is much less depressed.
E. Atlantic. 66 µm l, 72 µm w.

Protoperidinium grande (Kofoid) Balech
A very large species which is rather like a much extended *P. depressum*. Found in warmer
oceanic waters.
E. Atlantic. 140 μm l, 94 μm w.

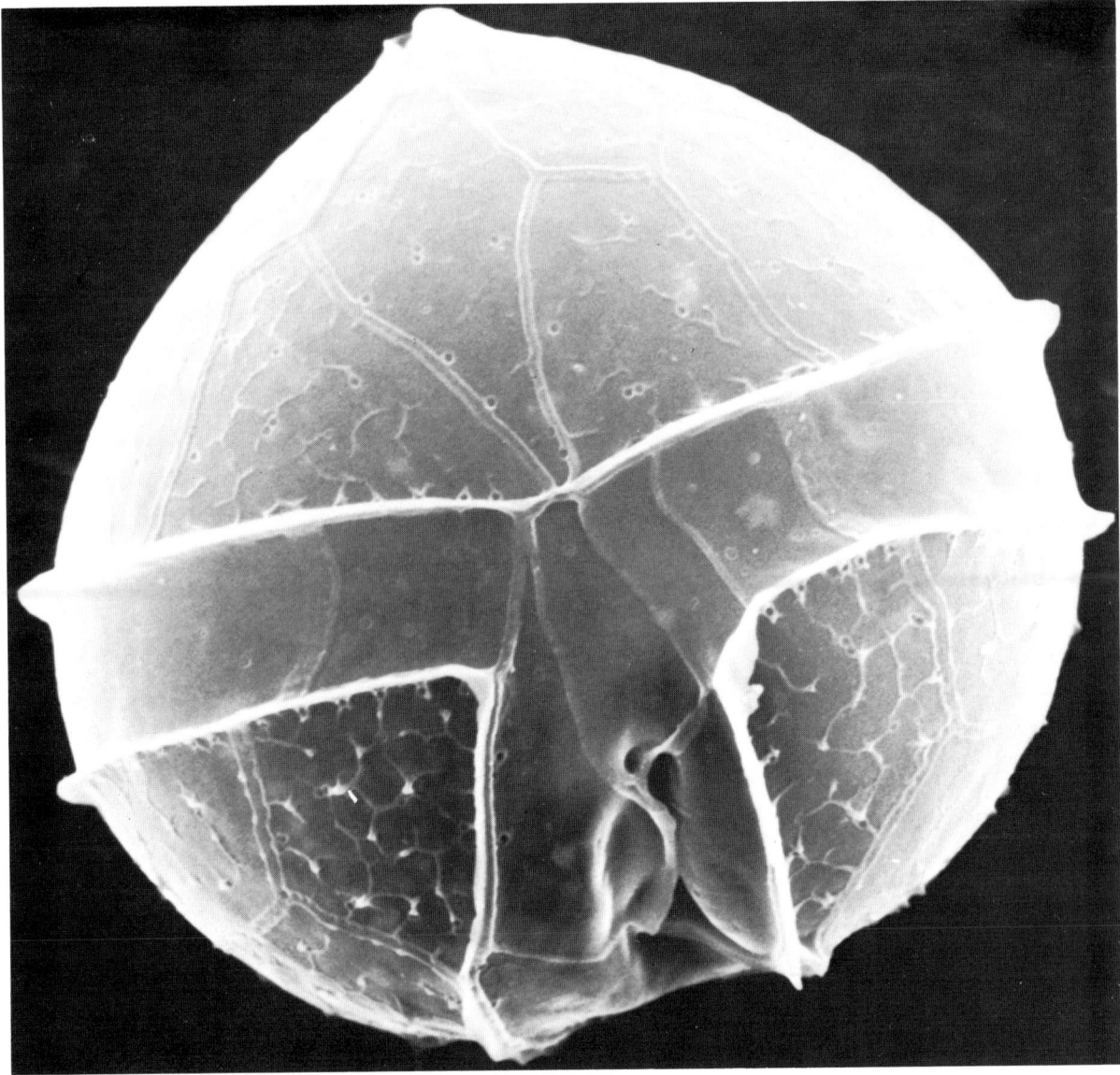

Protoperidinium nudum (Meunier) Balech
An ortho-*Protoperidinium* with a simple shape and very delicate ornamentation. In the
centre several of the sulcal plates and the flagellar pores can be clearly observed.
Atlantic, W. of Ireland. 25 μm l, 25 μm w.

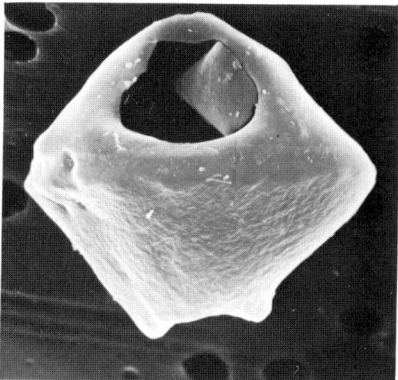

Protoperidinium leonis (Pavillard)
Balech
A distinctively straight-sided species with
a slightly offset median girdle. The ornamentation of the plates is somewhat varied, but often involves longitudinal
ridges as here. The cyst (small picture)
has a clearly peridinoid shape, is smooth,
and has a four-sided archeopyle.
North Sea. 70 μm l, 65 μm w.

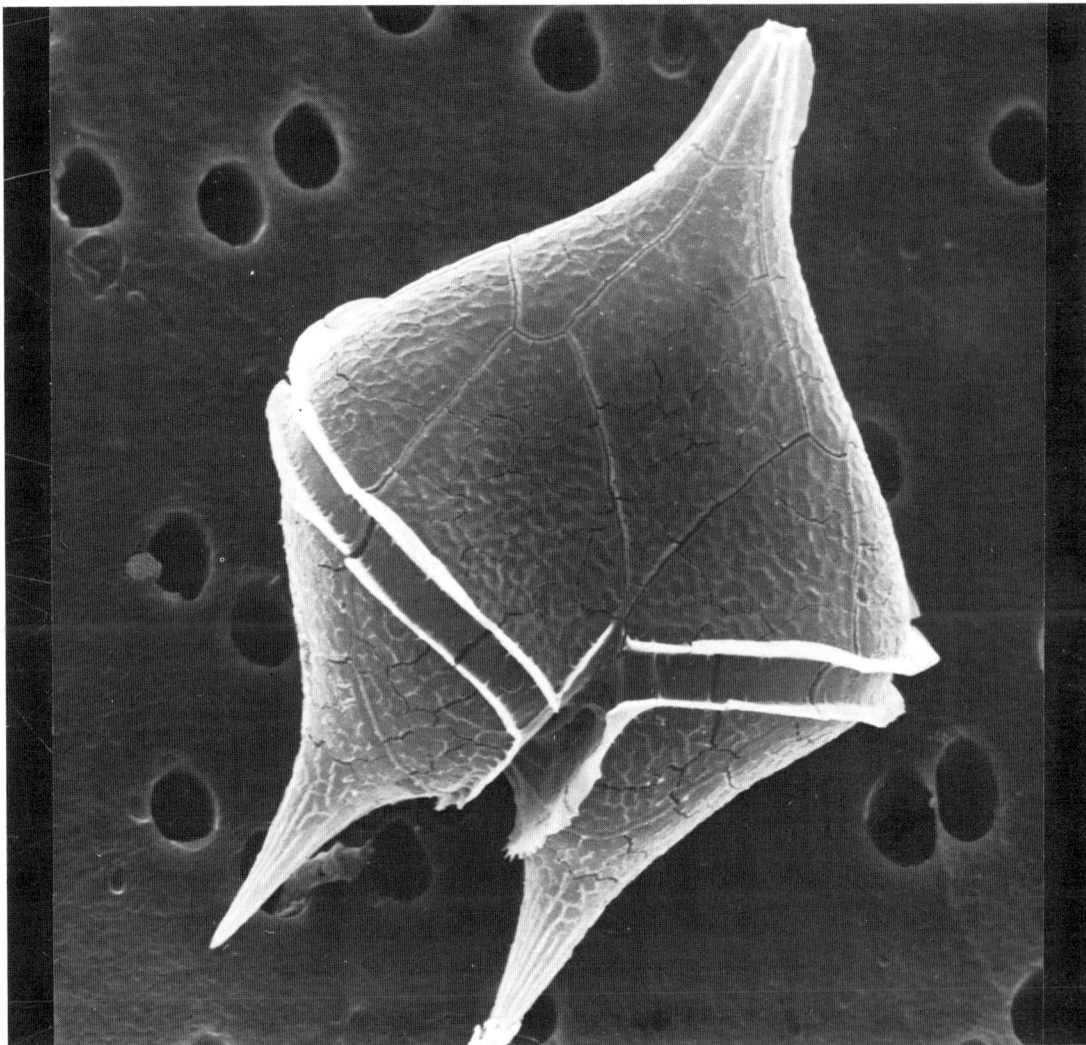

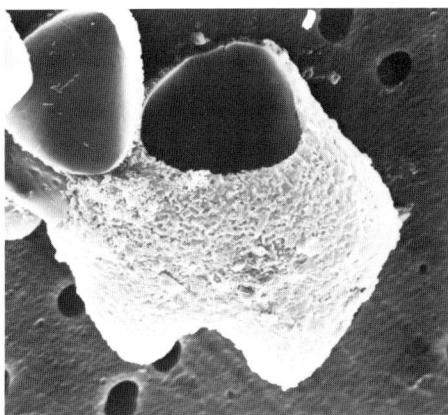

Protoperidinium oblongum (Aurivil-
lius) Parke & Dodge
A large ortho-*Protoperidinium* which is
not always easy to distinguish from *P.
claudicans* and *P. oceanicum*. It has a
distinctive cyst (small picture)
with a large excystment aperture (archeopyle).
North Sea. 105 μm 1, 70 μm w.

58

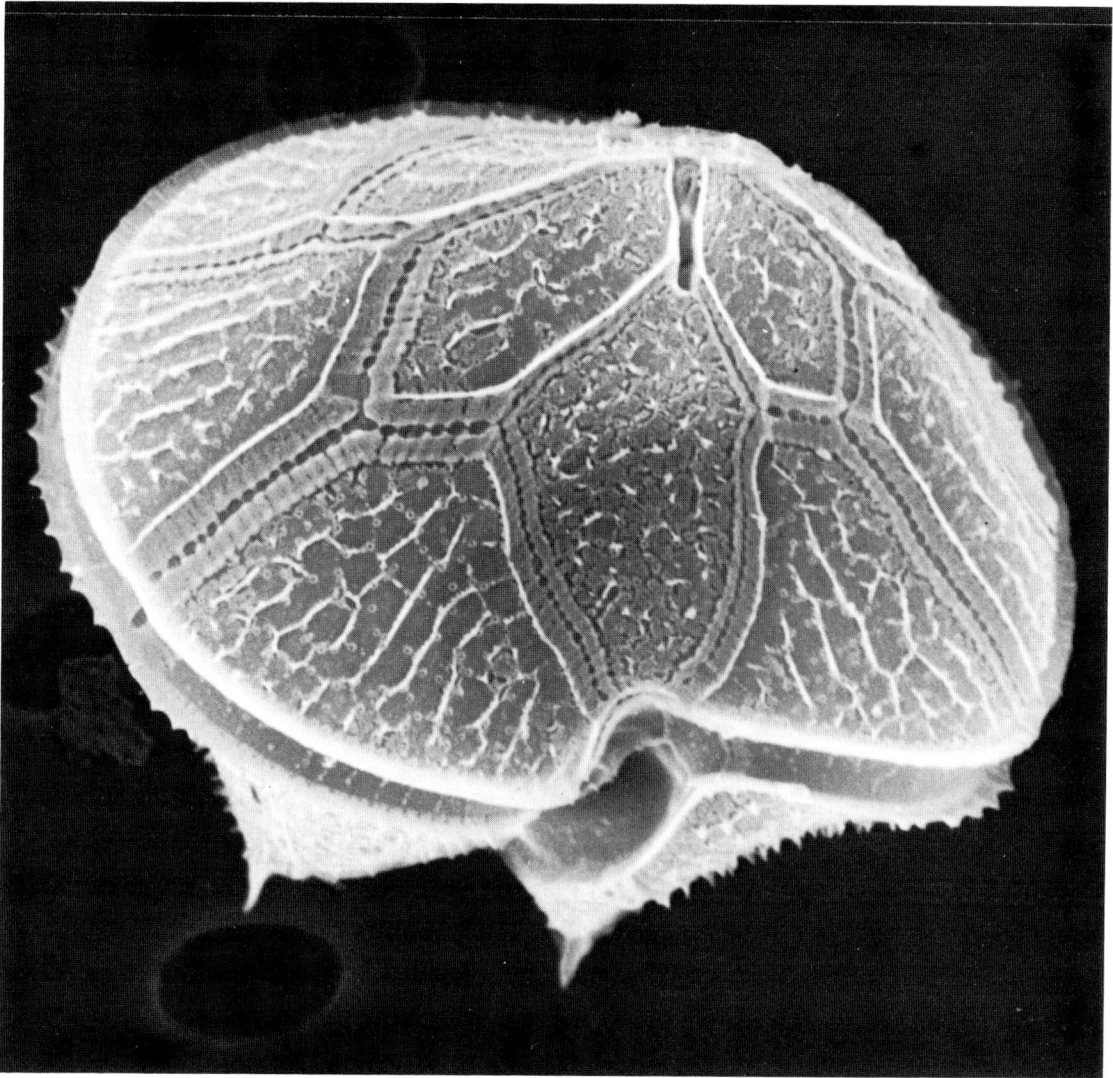

Protoperidinium obtusum (Karsten) Parke & Dodge
A species with a broad and blunt apex and a rather unusual ornamentation consisting of radial ridges. Note the ortho (four-sided) first apical plate which is clearly visible in this oblique apical view.
North Sea. 60 μm w.

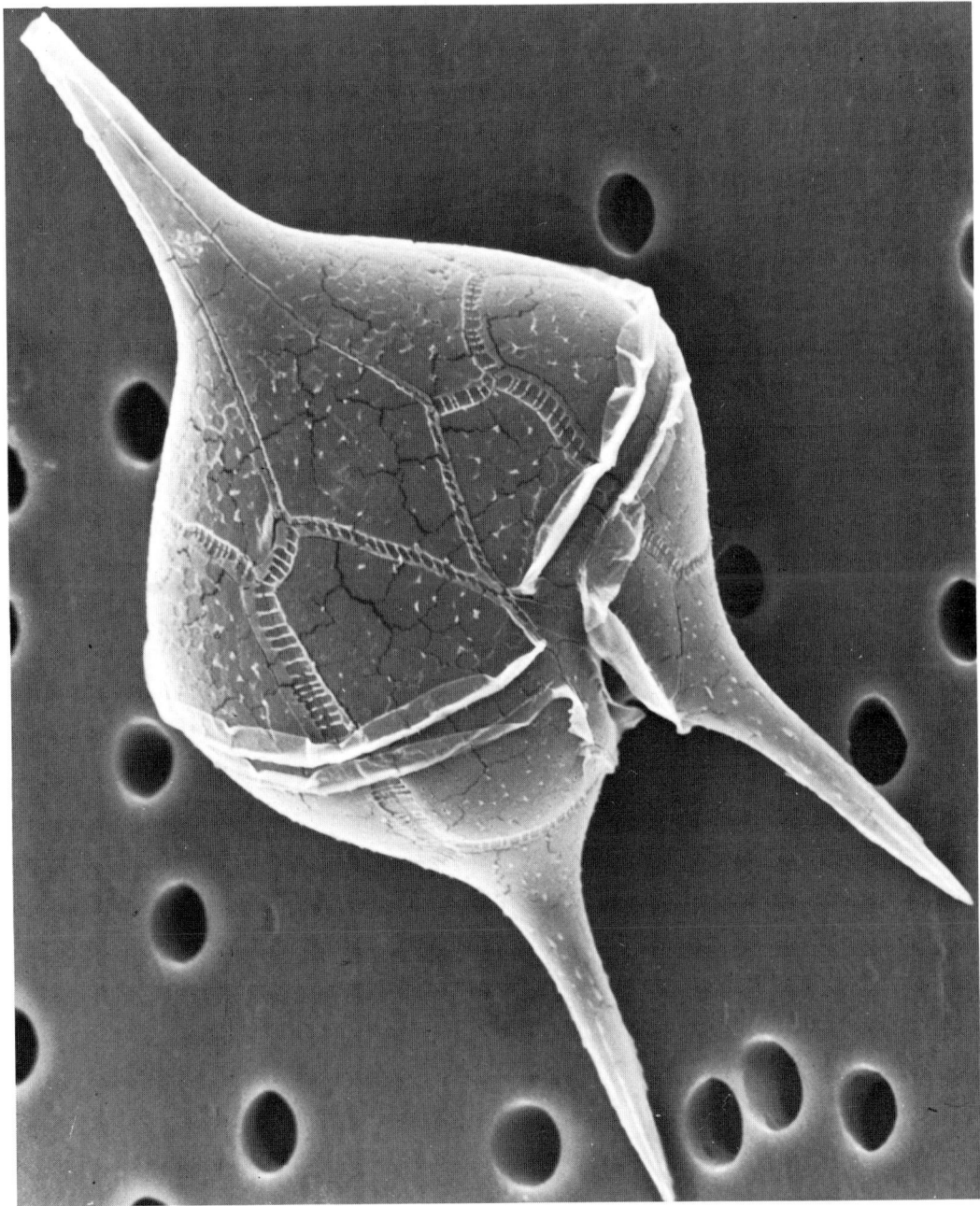

Protoperidinium oceanicum (Vanhöffen) Balech

A very large warm-water species which, as its name implies, is found in oceanic plankton.
It has an ortho first apical plate and rather smooth plate surfaces. As with *P. grande*, the
apical and antapical horns are much extended.
Tropical E. Atlantic. 160 µm l, 78 µm w.

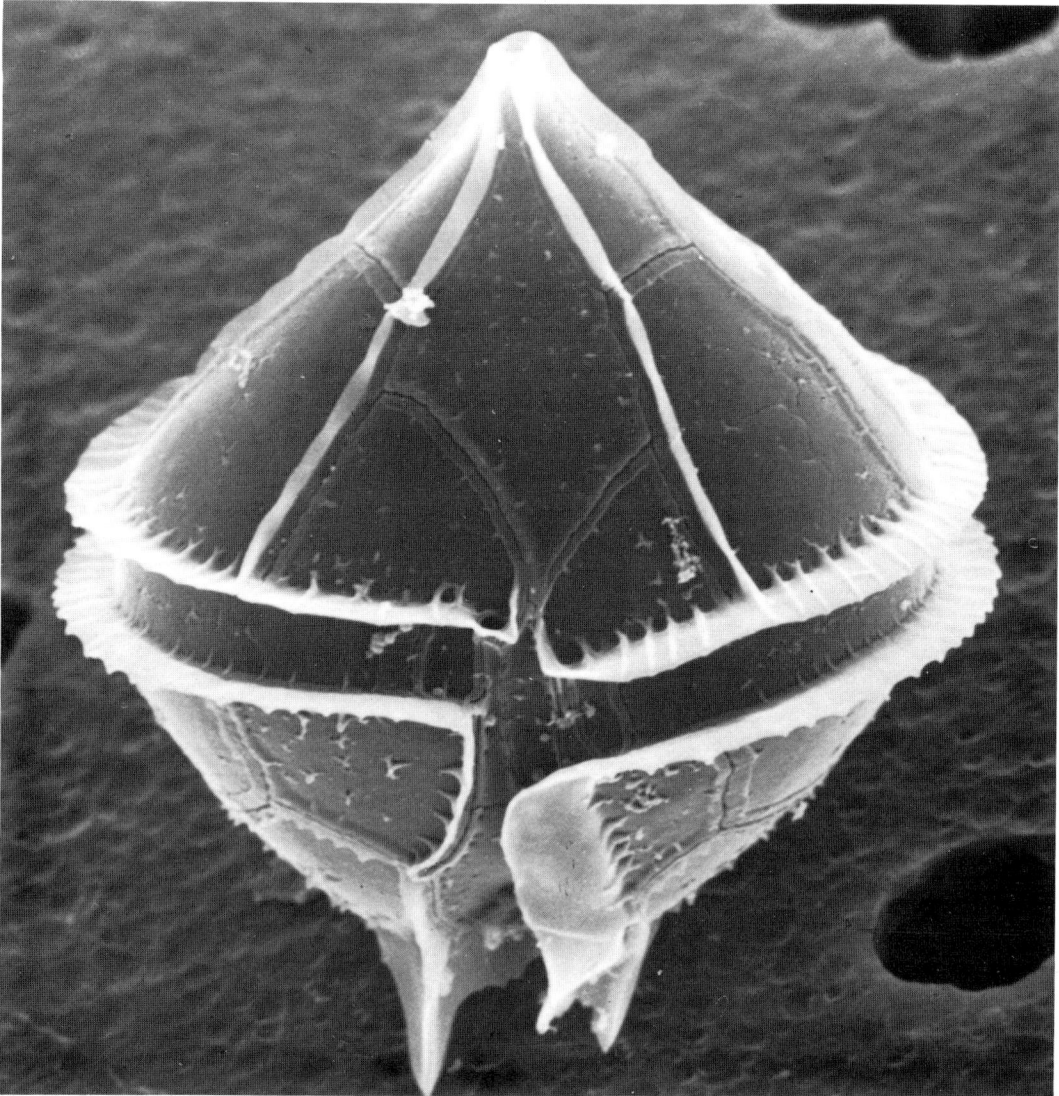

Protoperidinium pallidum (Ostenfeld) Balech
A para-*Protoperidinium* in which the first apical plate is six-sided. The sulcal wing is prominent in this species. Note the smoothness of the thecal plates. Common around the British Isles. North Sea. 53 μm l, 46 μm w.

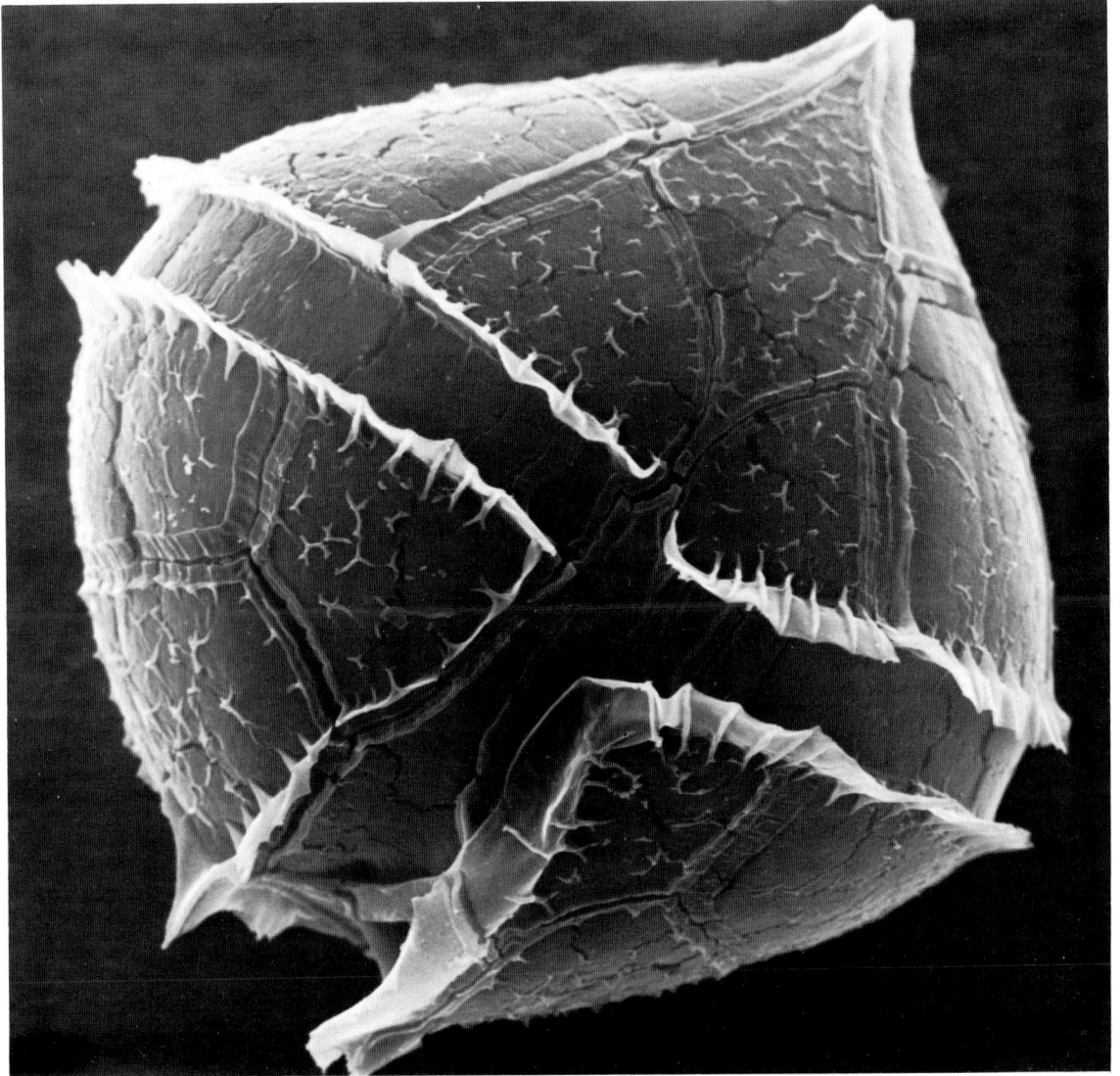

Protoperidinium pellucidum Bergh
A meta-*Protoperidinium* (five-sided first apical plate) which has a characteristic antapical projection from the sulcal wing. It is somewhat smaller than *P. pallidum* but is easily confused with that species.
Atlantic. 45 μm l, 40 μm w.

Protoperidinium pentagonum (Gran)
Balech
A large species in which the ventral face
is concave and the girdle displaced. The
cell has a five-sided appearance but the
first apical is clearly four-sided. The
spiny cyst (small picture) mimics the
shape of the cell and has a clear girdle
and sulcus.
North Sea. 90 μm l, 105 μm w.

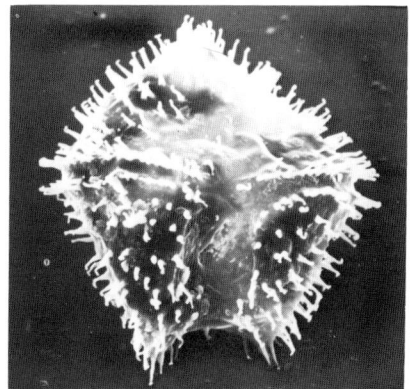

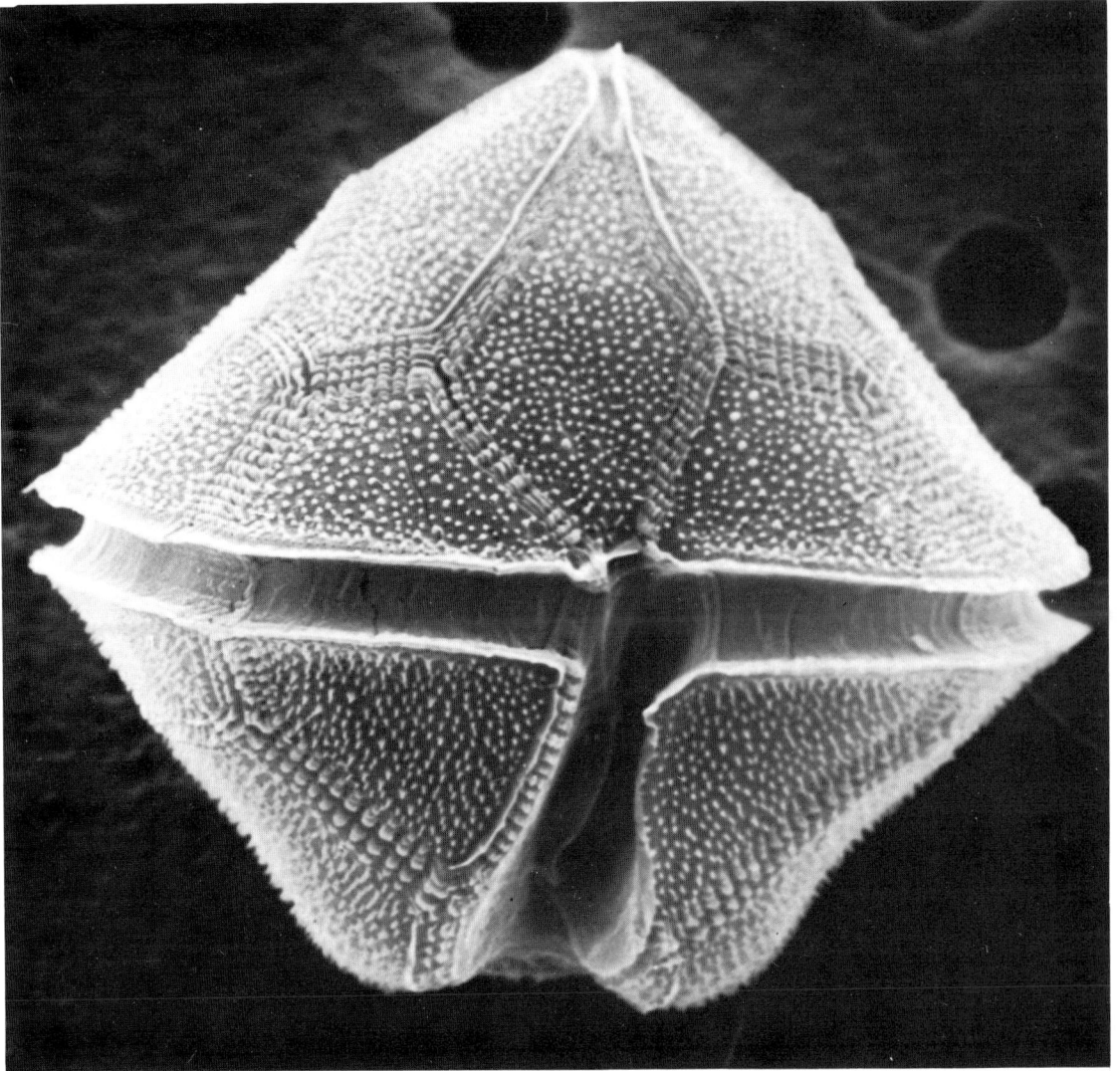

Protoperidinium punctulatum (Paulsen) Balech
A squarish ortho-*Protoperidinium* in which the thecal plates are covered by small pimples.
The name comes from the fact that in the light microscope the pimples can appear to be
pores or punctae.
S. Brittany. 45 µm l, 50 µm w.

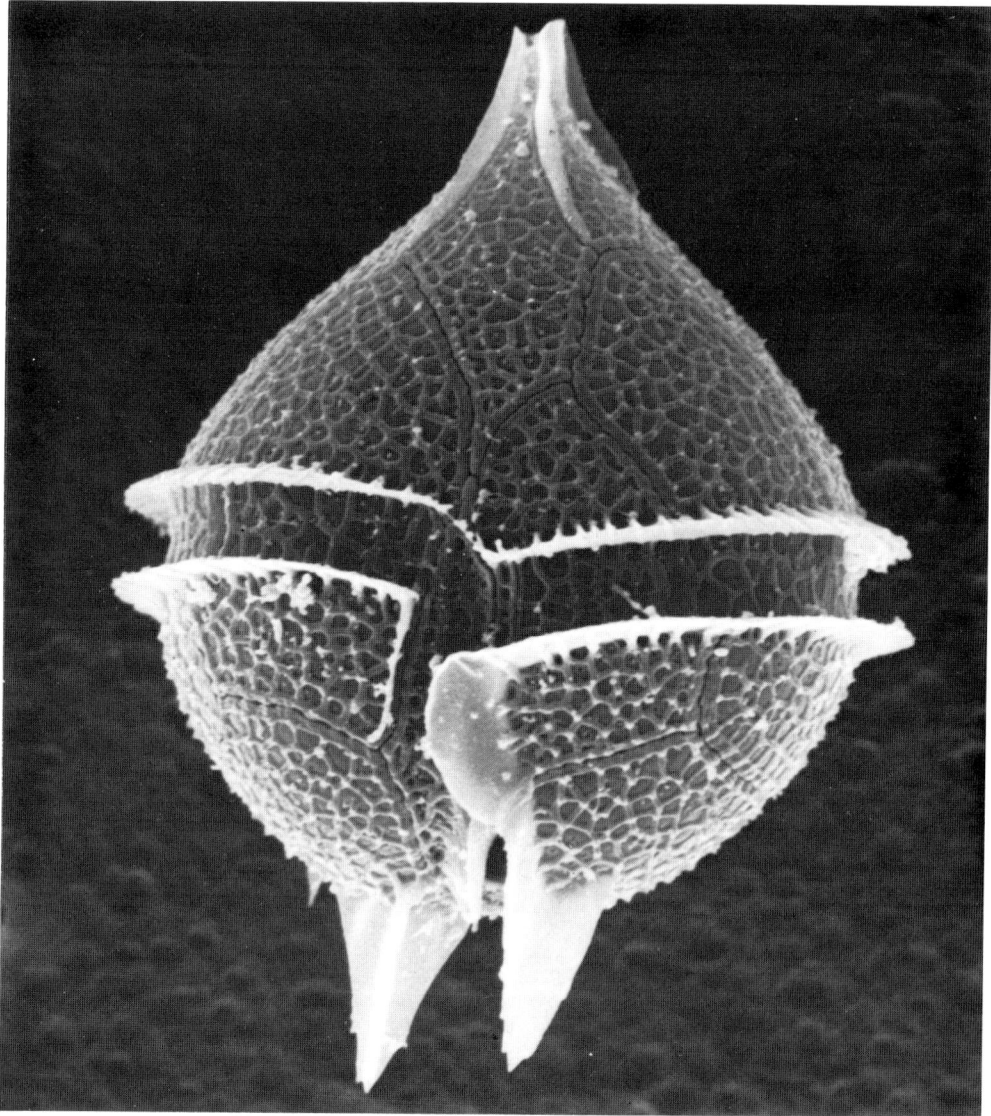

Protoperidinium steinii (Jørgensen) Balech
A rounded, top-shaped, species with only a short sulcal wing but long winged antapical
horns. The whole theca is covered with a regular reticulation pattern.
North Sea. 55 µm l, 40 µm w.

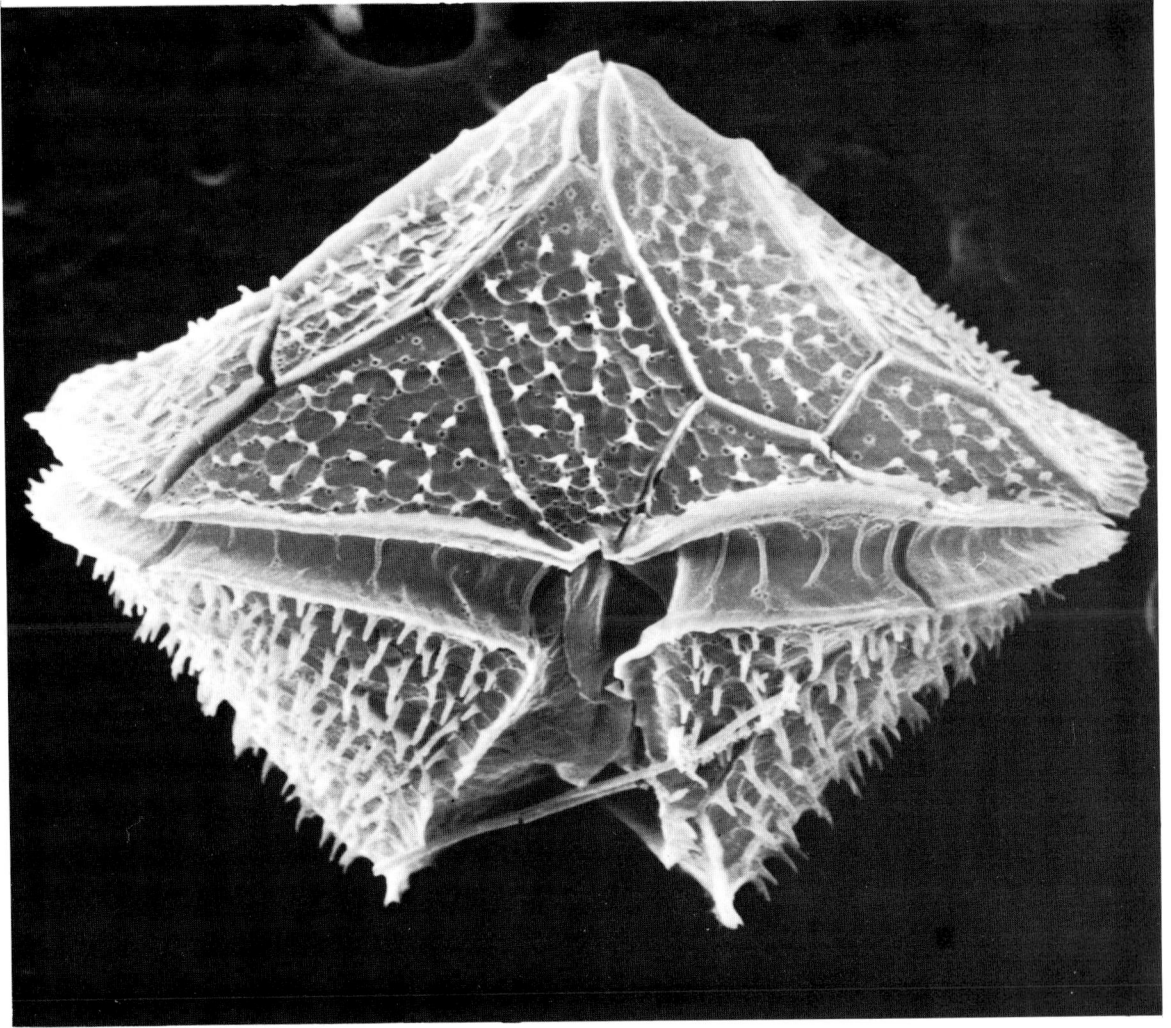

Protoperidinium subinerme (Paulsen) Loeblich
A squarish dinoflagellate which is wider than it is long. Note the straight sides and ortho
first apical plate. On the hypotheca the reticulated ornamentations are extended to form
small spines.
S. Brittany. 44 µm l, 54 µm w.

66

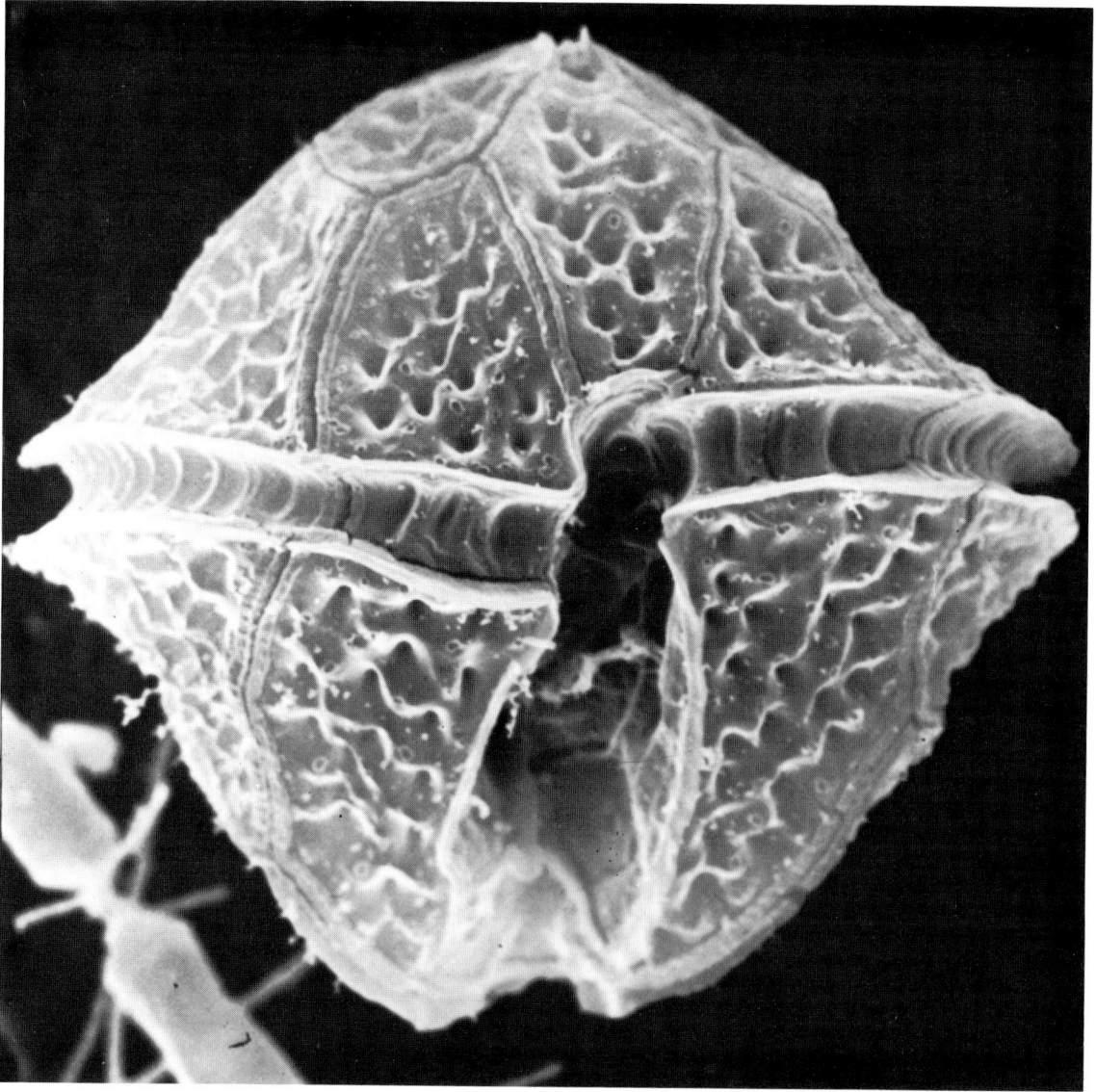

Protoperidinium thorianum (Paulsen) Balech
An ortho-*Protoperidinium* with a compact shape (i.e. no horns), an offset girdle and a rugose
surface to the thecal plates. Found all around the British Isles.
E. Atlantic. 56 µm l, 58 µm w.

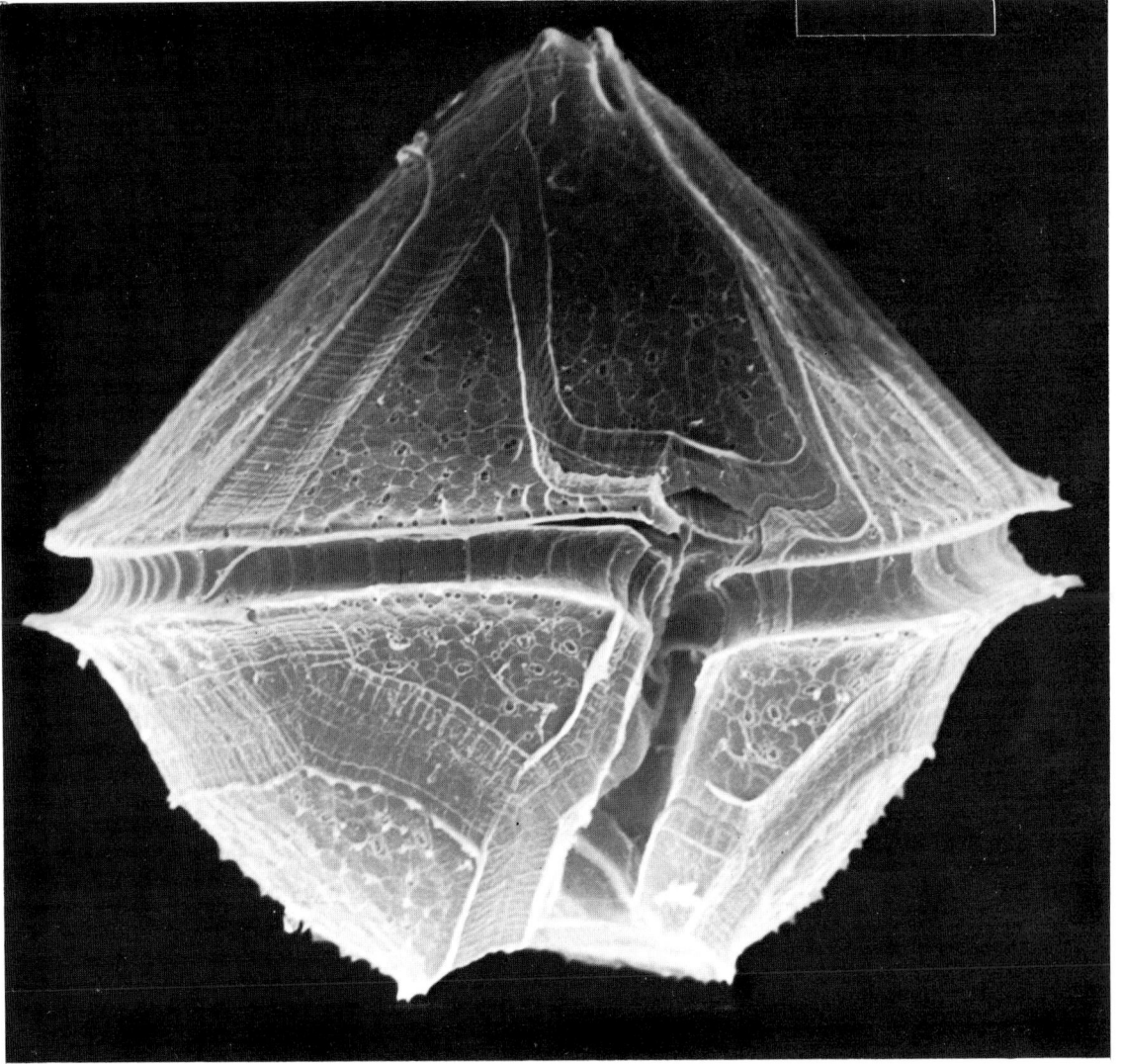

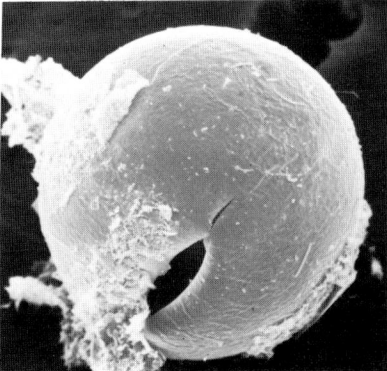

Protoperidinium thulesense (Balech)
Balech
An unusual species with an almost rec-
tangular first apical plate. The epitheca is
straight sided and the hypotheca more
rounded. Note the broad sutural bands
in this specimen. There is a round
brown cyst (small picture) with a slit-like
excystment pore.
North Sea. 60 μm l, 62 μm w.

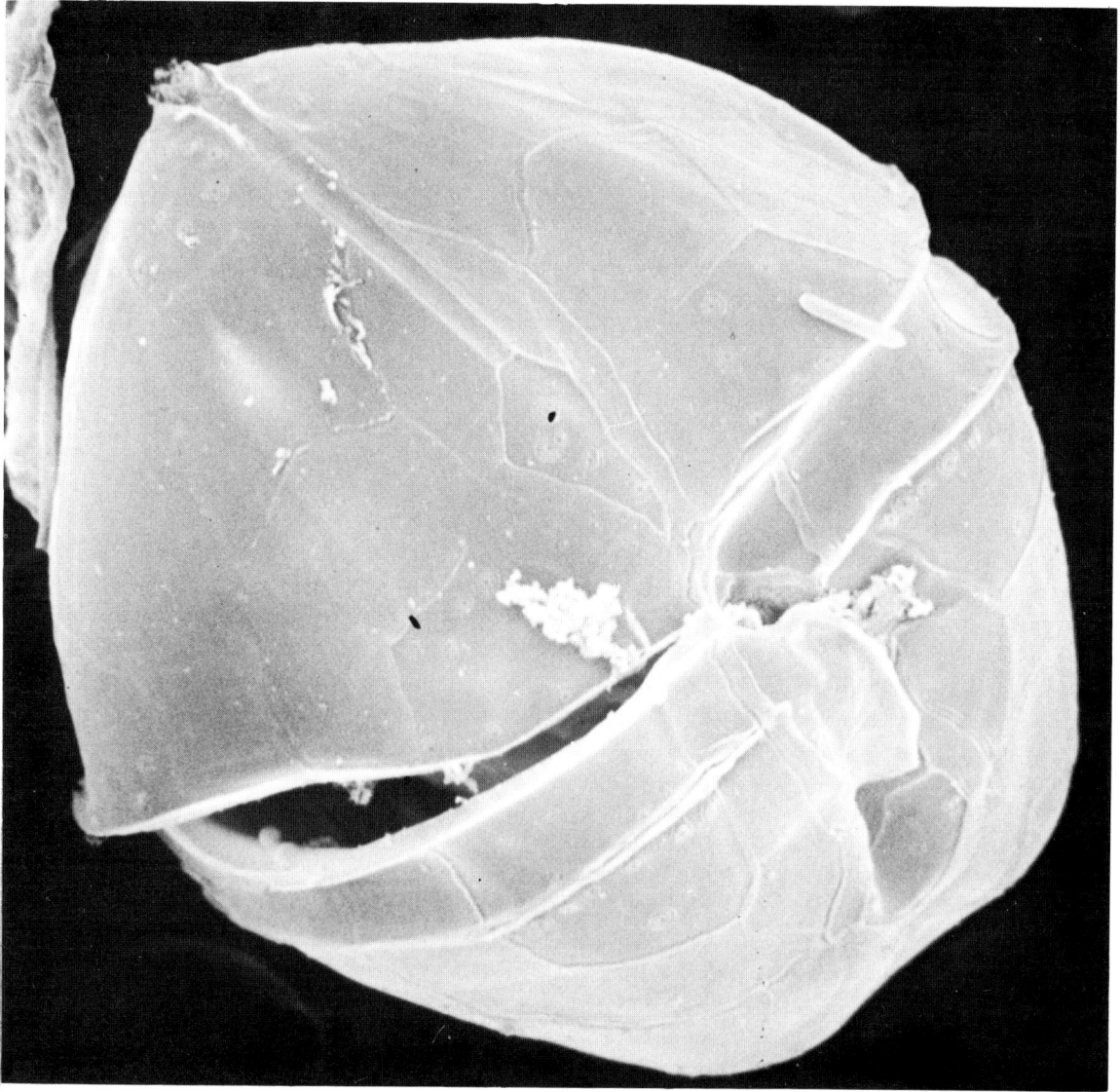

Scrippsiella faeroense (Paulsen) Balech
& Soares
A small pyriform dinoflagellate which by
light microscopy is easily confused with
other species of *Scrippsiella*. Note the
very short first apical plate. There are
rather unusual concentric patterns on the
thecal plates (small picture).
Irish Sea. 32 μm l, 25 μm w.

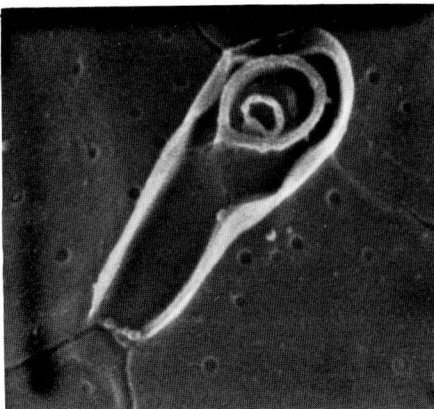

Scrippsiella trochoidea (Stein) Loeblich
This small and relatively undistinguished
photosynthetic dinoflagellate is, with its
relatives, fairly common in neritic waters.
Note the slightly offset girdle and plain
thecal plates. The lanceolate first apical
plate stretches from the sulcus to the
apical pore (small picture). This species
forms a calcareous cyst.
E. Atlantic. 20 μm l, 16 μm w.

70

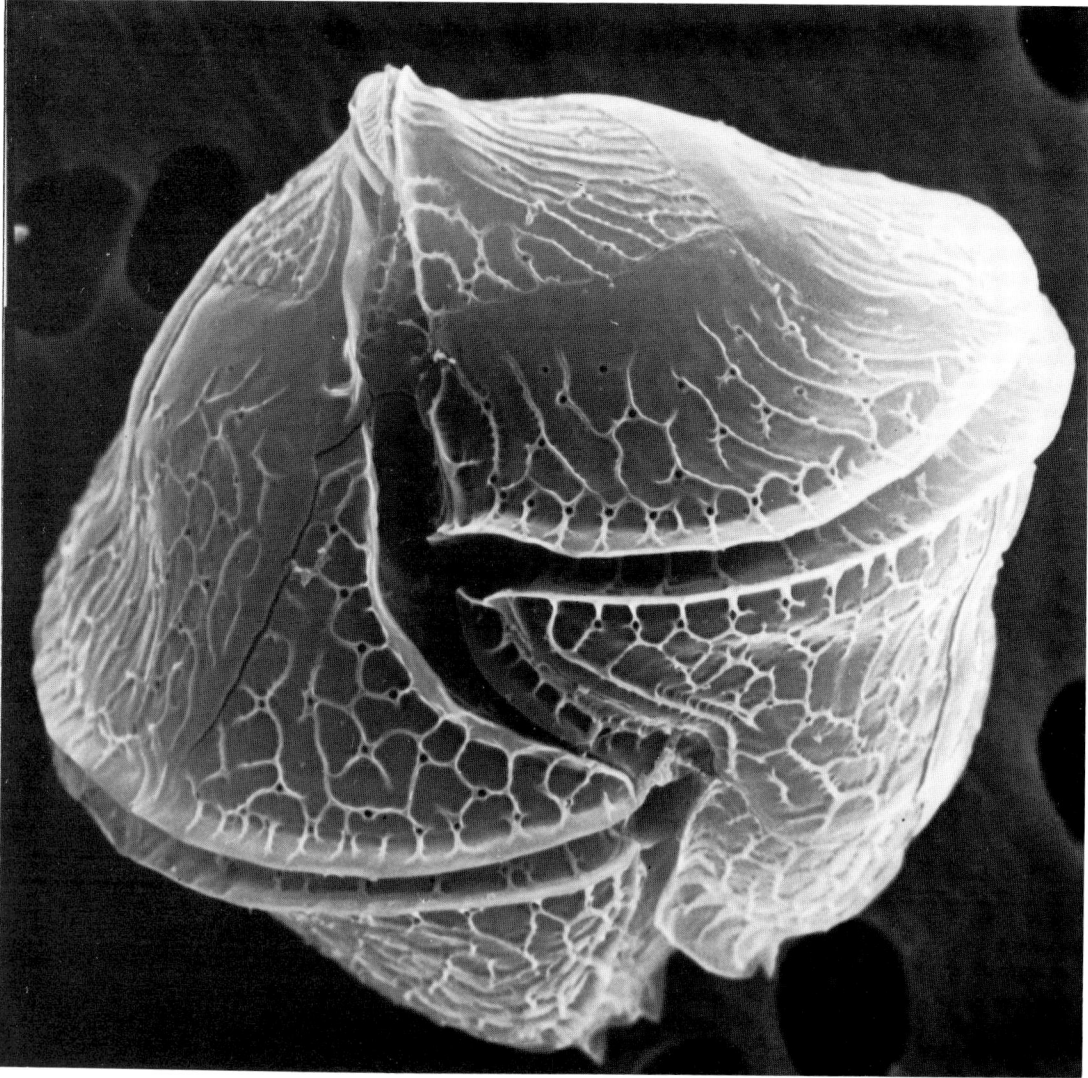

Gonyaulax alaskensis Kofoid
A large and rather rhomboidal species with a delicate reticulate patterning over the theca. Originally described from the Pacific Ocean but also found in the North Sea. The typical plate pattern in this genus is: Po, 3', 2a, 6'', c, s, 5''', 1p, 1''''. The small picture shows the apical plates.
North Sea. 60 μm l, 60 μm w.

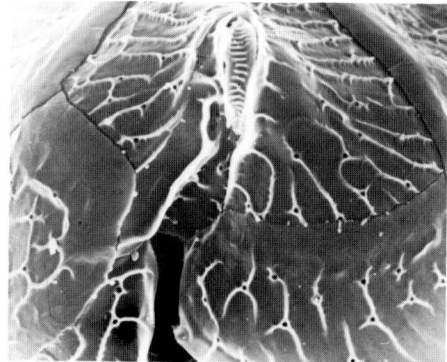

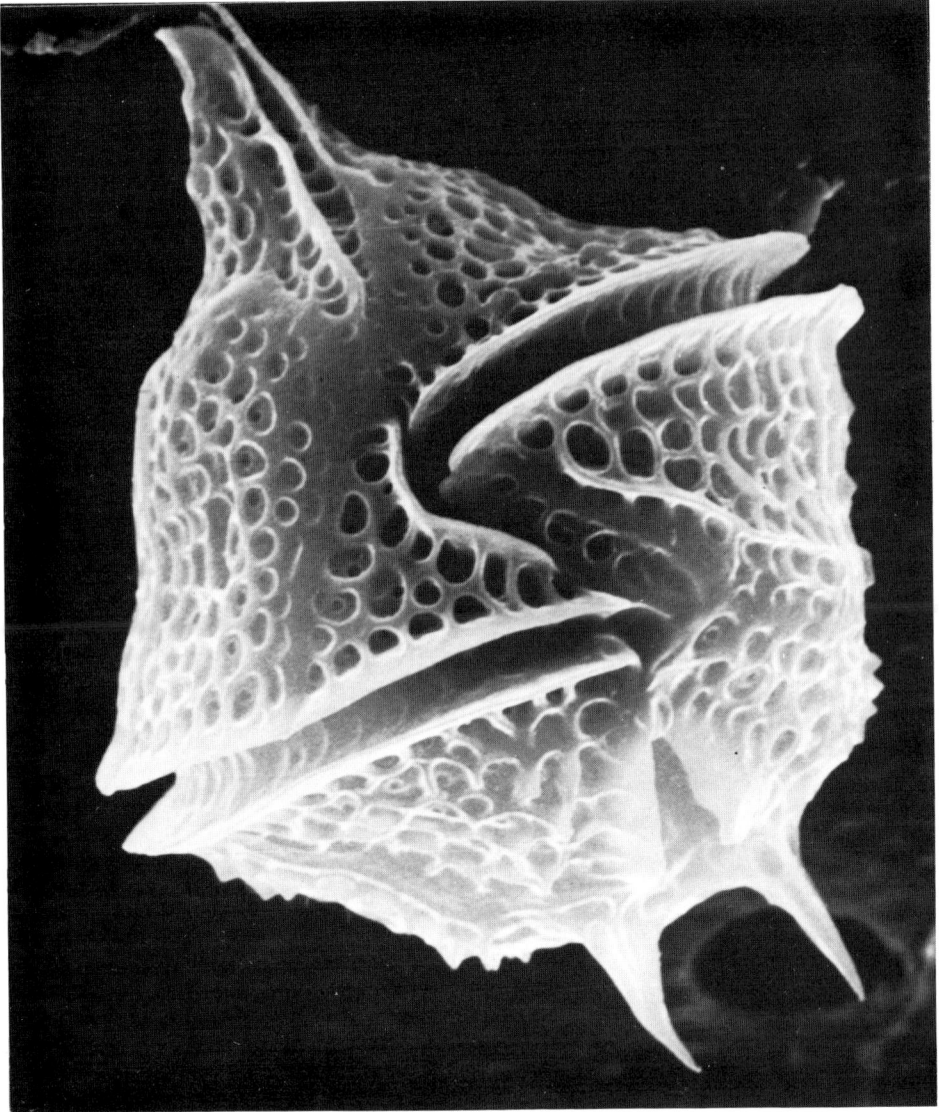

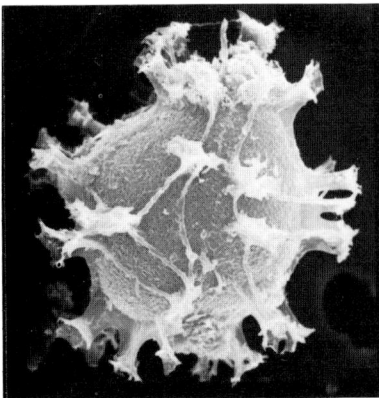

Gonyaulax digitale (Pouchet) Kofoid
This fairly common temperate species is
characterised by its robust theca, distinctive apical horn and two stout antapical
spines. The cyst stage (small picture)
called *Spiniferites bentori* has rather ela-
borate sutural processes and the girdle is
quite clearly outlined.
North Sea. 60 μm l, 45 μm w.

Gonyaulax jollifei Murray & Whitting
A very strikingly shaped *Gonyaulax* with thick thecal plates and strong horns. An oceanic organism, this is probably the same species as *G. fusiformis* which has been described from the Pacific Ocean.
E. Atlantic. 55 μm l, 30 μm w.

Gonyaulax monospina Rampi

A relatively rare species which has the characteristic *Gonyaulax* offset girdle and thecal tabulation. The cell is much more rounded than in most other species and only has one small antapical spine.

E. Atlantic. 33 μm l, 27 μm w.

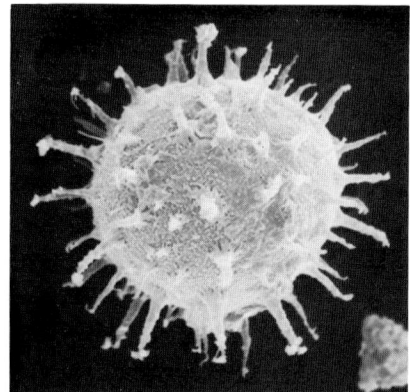

Gonyaulax polyedra Stein
An angular species which has thick thecal plates with ridges following the sutures. The cyst (small picture) is spherical with numerous rough projections and is also known as *Lingulodinium machaerophorum*
W. Scotland. 30 μm l, 30 μm w.

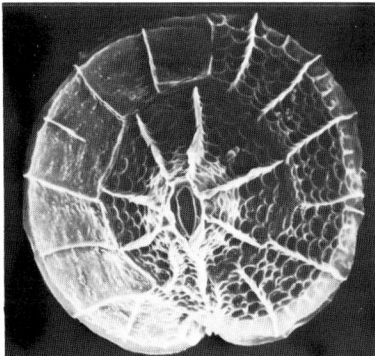

Gonyaulax polygramma Stein
A common oceanic species with characteristic strong longitudinal ridges on the thecal plates. The girdle is not so strongly displaced as in many *Gonyaulax* spp. The plate thickness and patterning can vary greatly, as is illustrated by the apical view of a recently divided cell (small picture).
E. Atlantic. 50 μm l, 40 μm w.

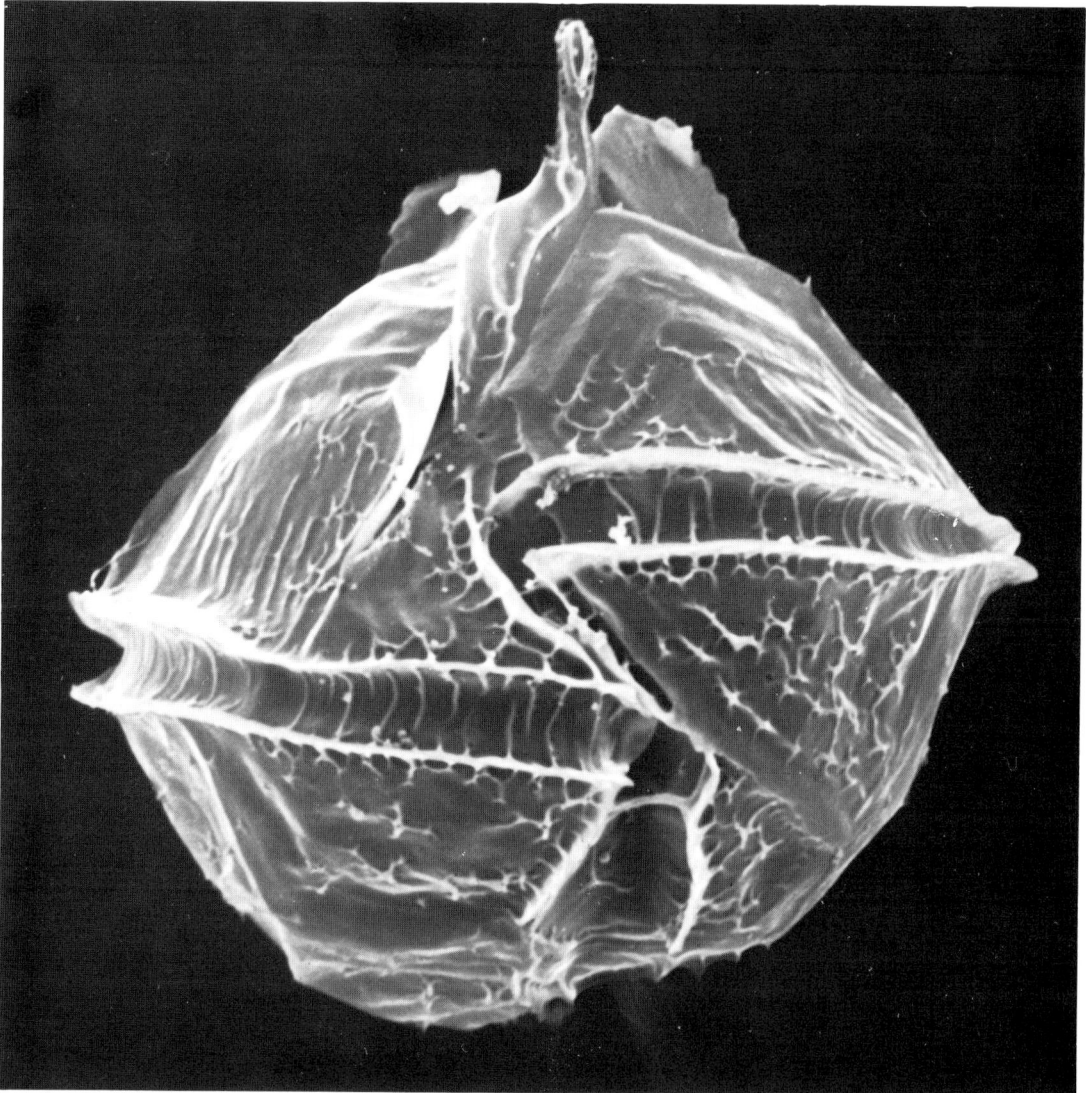

Gonyaulax scrippsae Kofoid
A small and rather undistinguished
Gonyaulax from oceanic waters. In this
specimen the thecal plates are very
lightly ornamented and the girdle is
characteristically displaced. The cyst
(small picture) is ovoid with numerous
forked sutural processes and is known as
Spiniferites bulloideus.
E. Atlantic. 40 μm l, 40 μm w.

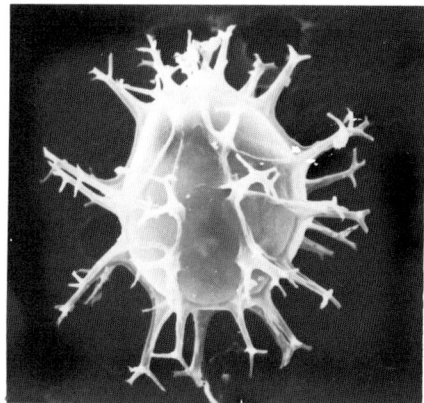

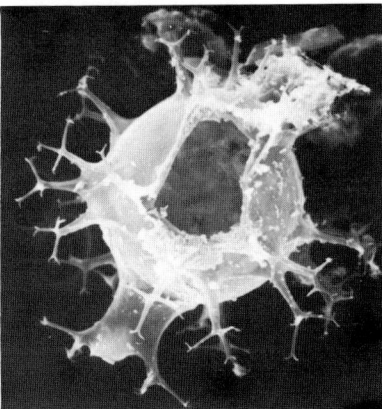

Gonyaulax spinifera (Claparédè &
Lachmann) Diesing
The type species of the genus but a real
problem regarding its correct identifica-
tion since it has been confused with *G.
digitale* and other species. It has a fairly
delicate theca with several spines at the
posterior end and a very big displacement of the girdle. Several types of cyst
are reported and one, *Spiniferites mirabi-
lis* is shown.
North Sea. 52 μm l, 44 μm w.

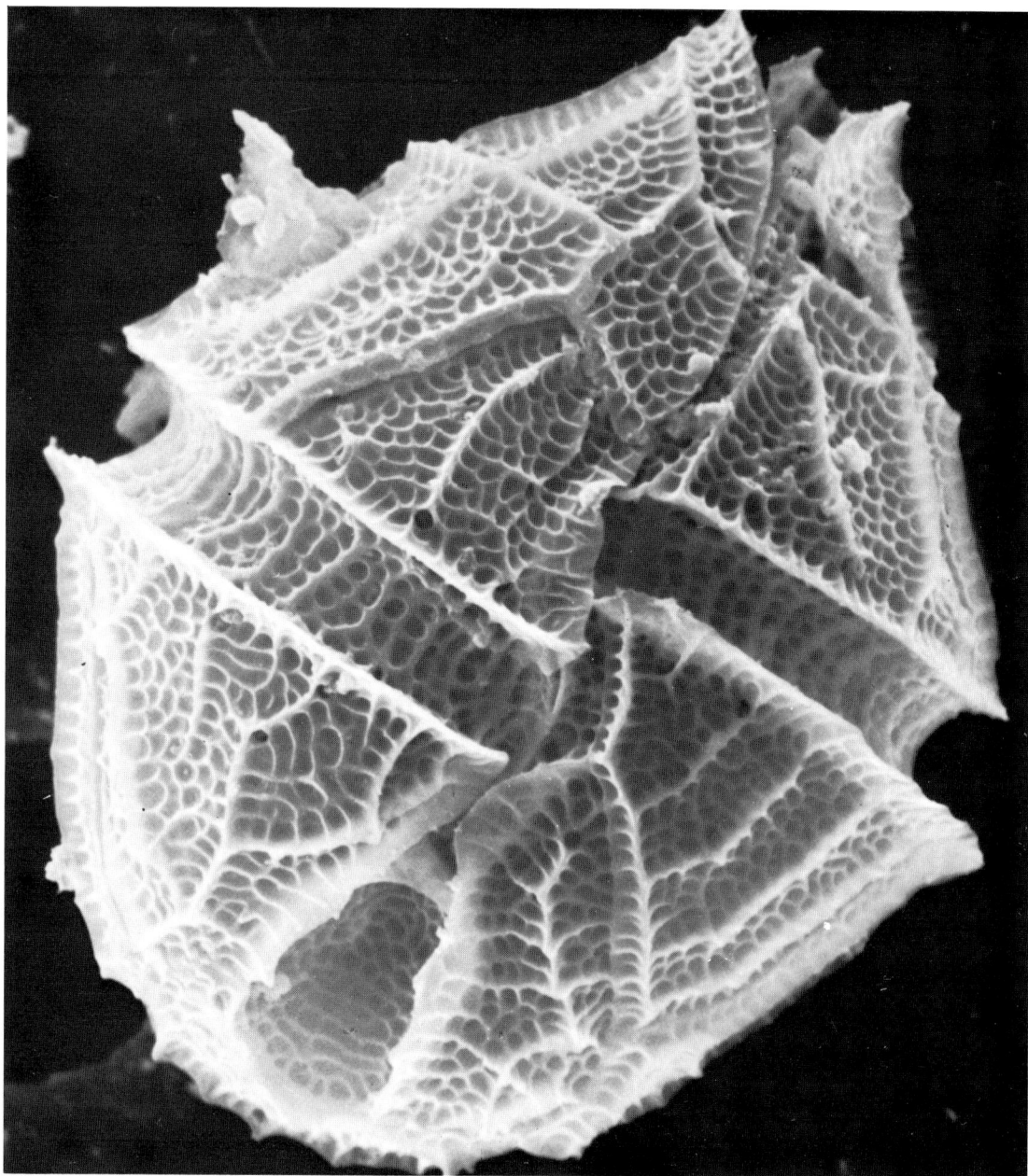

Gonyaulax striatum Mangin
A small oceanic species with a very heavily ornamented theca. This is almost certainly related to *G. polygramma* from which it differs by being smaller, differently patterned and having a very deeply recessed sulcus (small picture).
E. Atlantic. 30 μm l, 25 μm w.

Gonyaulax triacantha Jørgensen
A distinctively angular species with long apical horn and a varied number of antapical horns. Here in the SEM the thecal plates appear strongly marked but by light microscopy they are rather transparent. The epithecal plates differ from those of most *Gonyaulax* species.
W. Scotland. 50 μm l, 27 μm w.

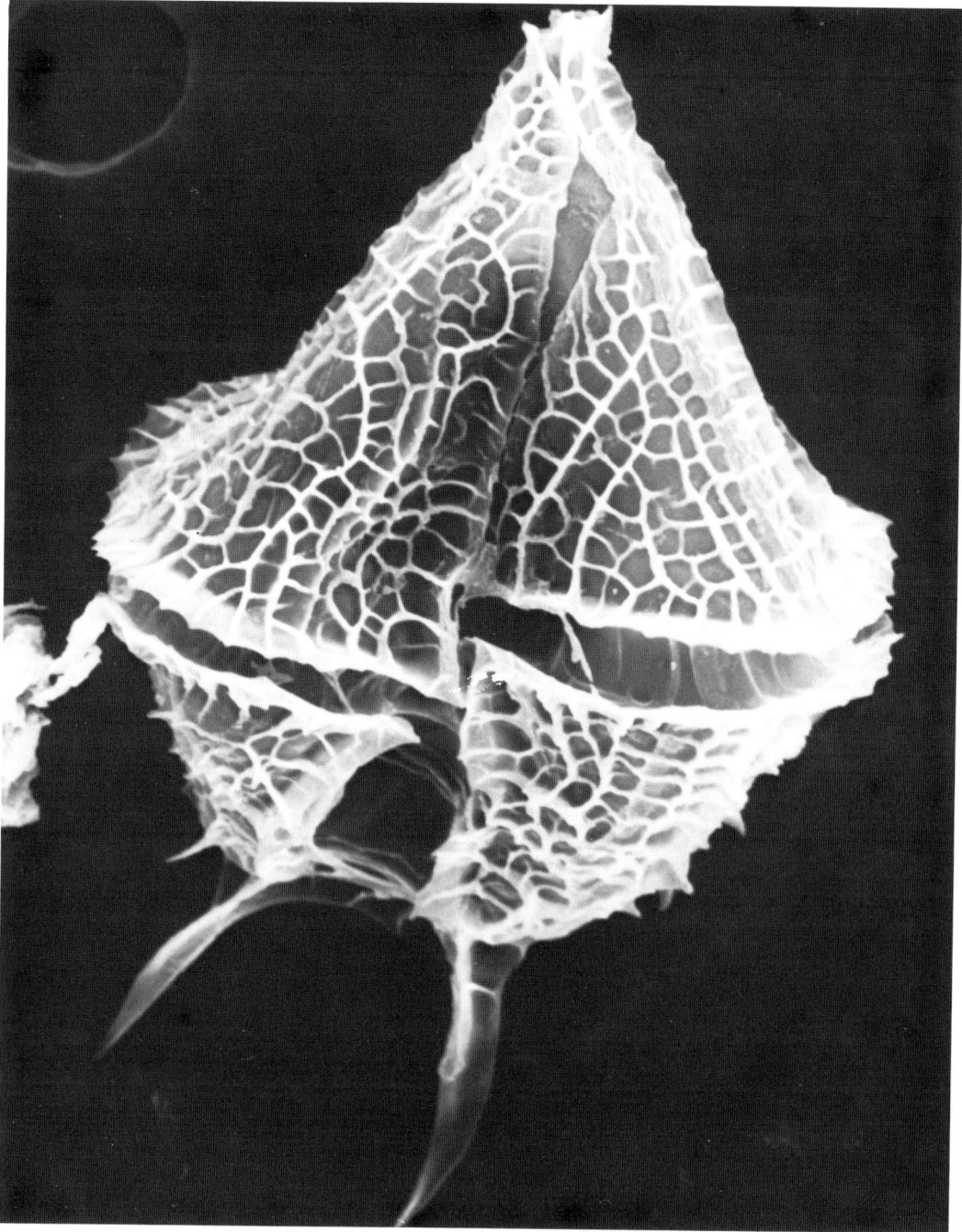

Gonyaulax verior Sournia

The combination of conical epitheca and two large antapical horns makes this species quite easy to recognise. The narrow first apical plate is reduced and unlike most other thecal plates is smooth and unornamented. A species of brackish and neritic waters.

North Sea. 50 μm l, 32 μm w.

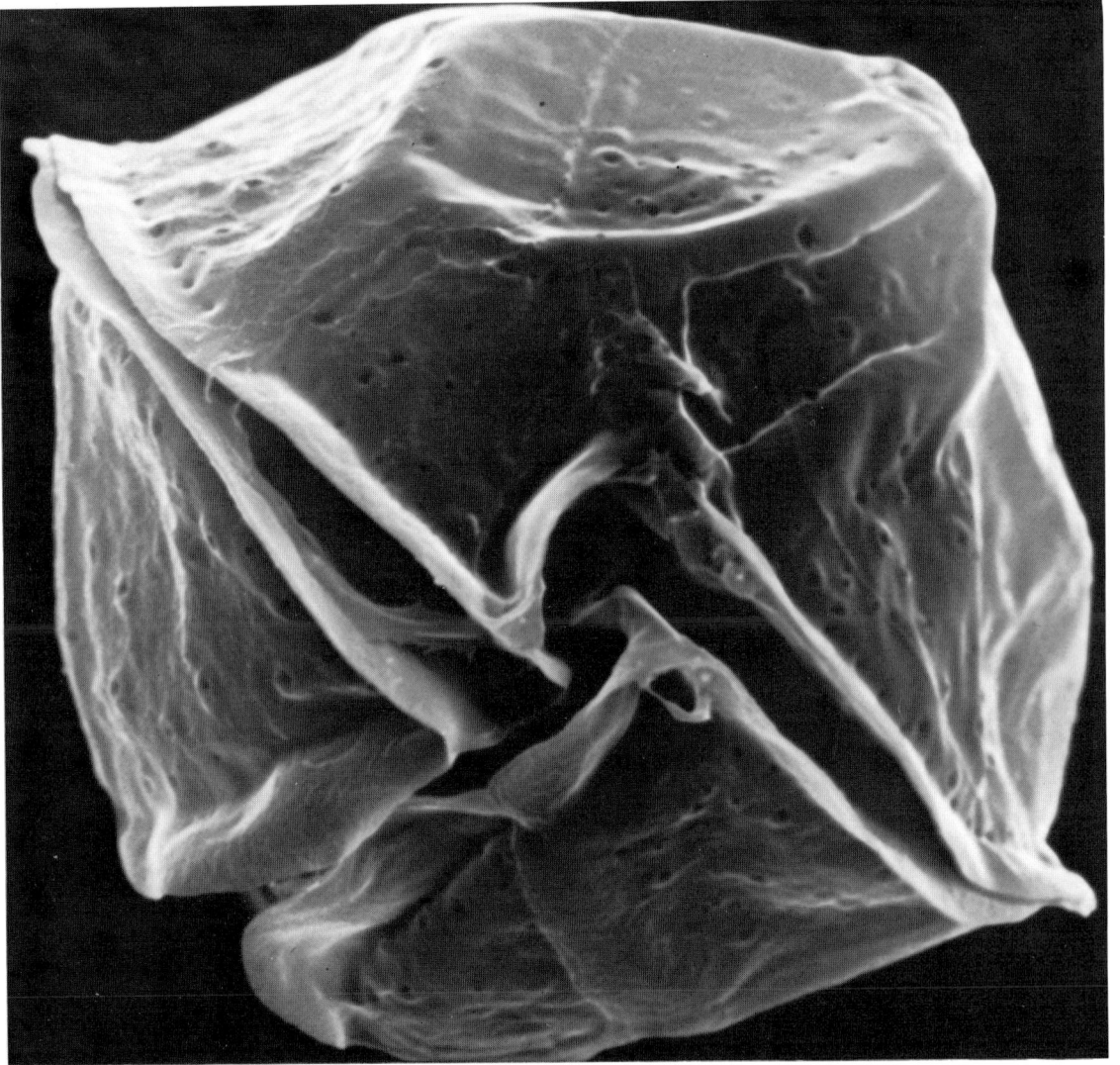

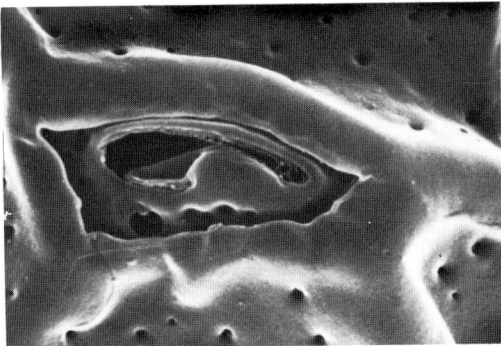

Protogonyaulax tamarensis (Lebour) Taylor

A small and rather undistinguished looking dinoflagellate from neritic (near coast) waters. It is, however, responsible for many of the occurrences of paralytic shellfish poisoning (PSP) which cause problems for shellfish farmers and eaters. Note the slightly offset girdle, smooth thecal plates, and curved apical pore (small picture).

North Sea. 30 μm l, 35 μm w.

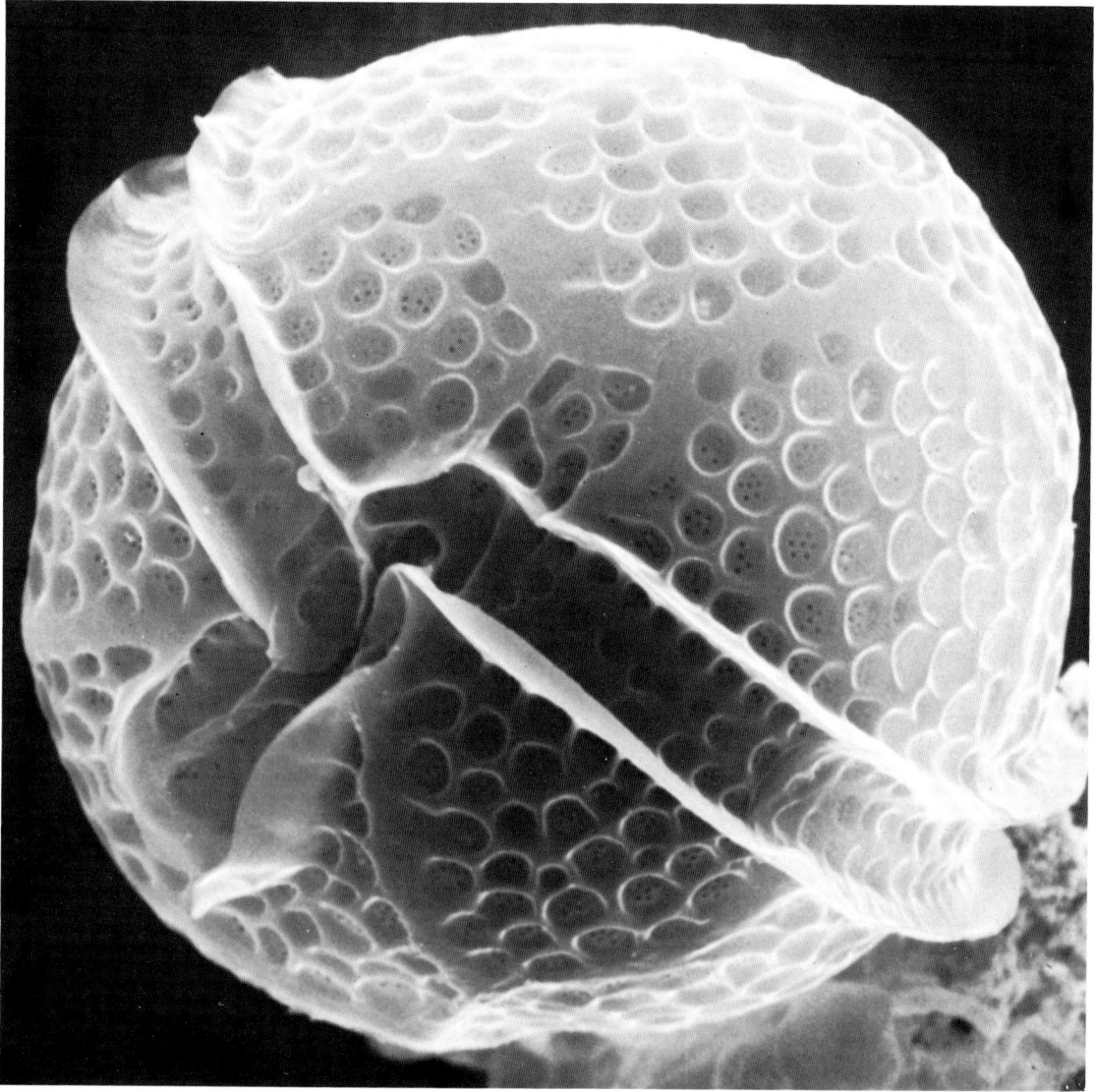

Peridiniella sphaeroidea Kofoid &
Michener
A rare and little-known species of oceanic waters. The plate pattern is obscured
by the very strong ornamentation of the
theca, each reticulation enclosing a cluster of small pores. The small picture
shows the apical pore. *Peridiniella* is closely related to the genus *Gonyaulax*.
E. Atlantic. 30 μm d.

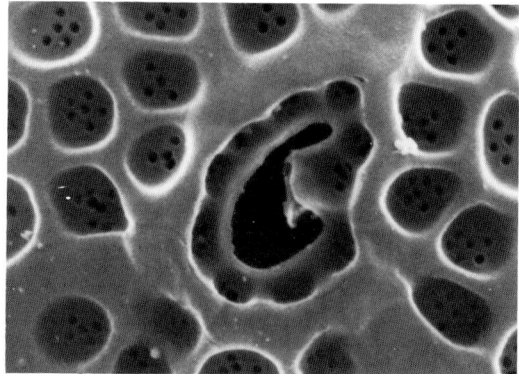

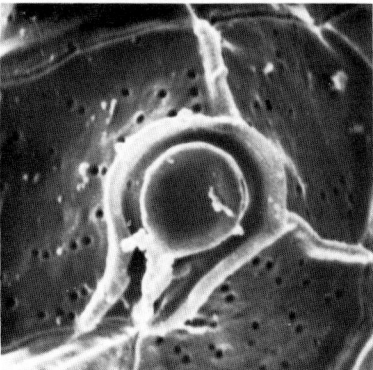

Amphidoma caudata Halldal
A small and inconspicuous photosynthetic dinoflagellate which is now known to be present around the north and west of the British Isles. The theca seems smooth and featureless, but it is divided into plates in the usual way. An antapical projection is typical of this genus. The small picture shows the apical pore.
Galway Bay, Ireland. 35 µm l, 30 µm w.

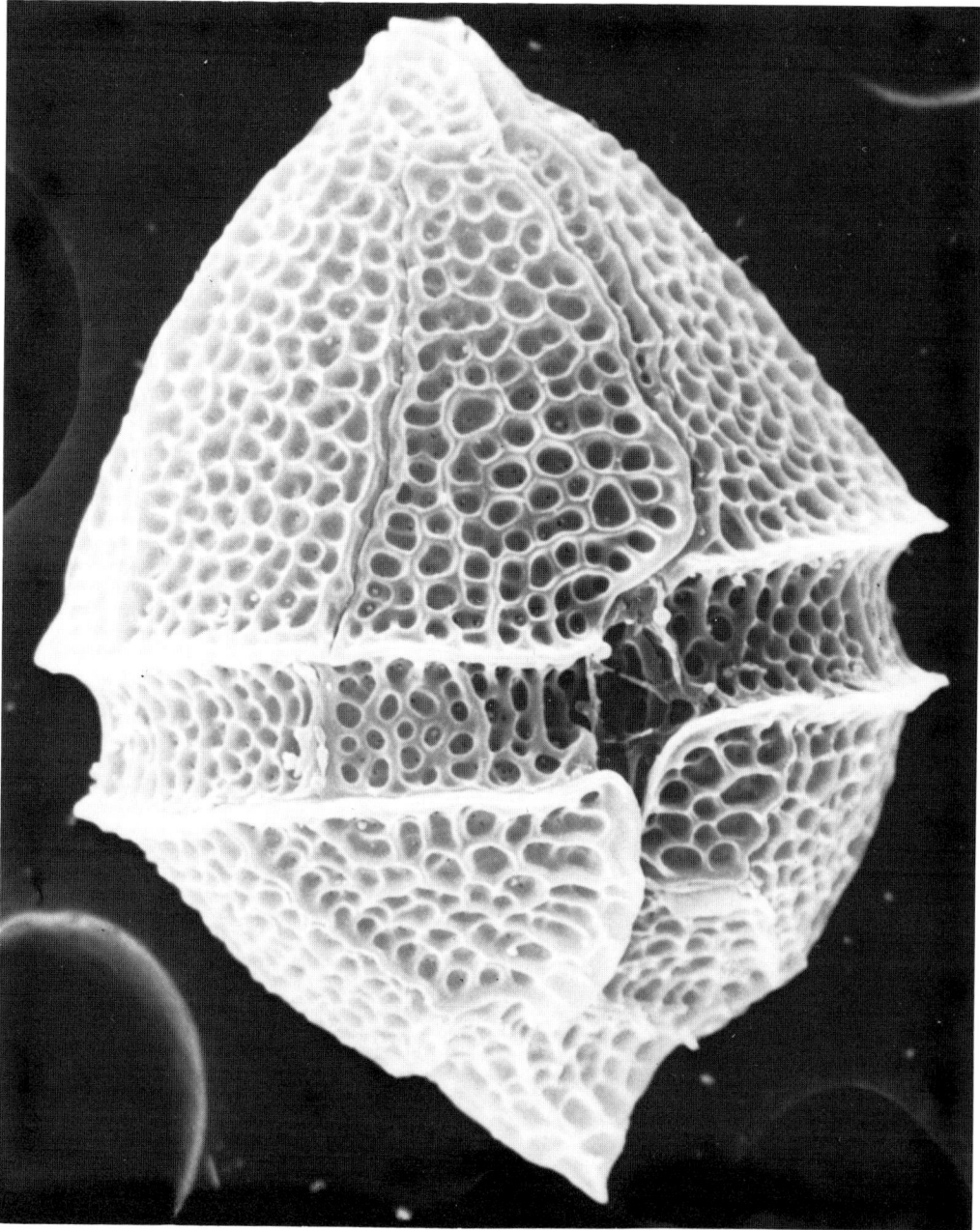

Amphidoma nucula Stein
A small and heavily armoured biconical
oceanic dinoflagellate. There is a distinctive pattern of thecal plates with large
pre-cingulars and a cluster of small
plates around the apex as shown in the
small picture of another, unidentified,
species.
E. Atlantic. 30 μm l, 22 μm w.

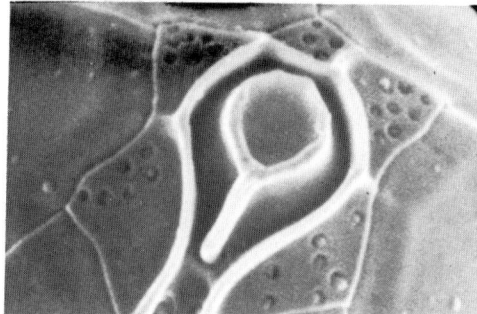

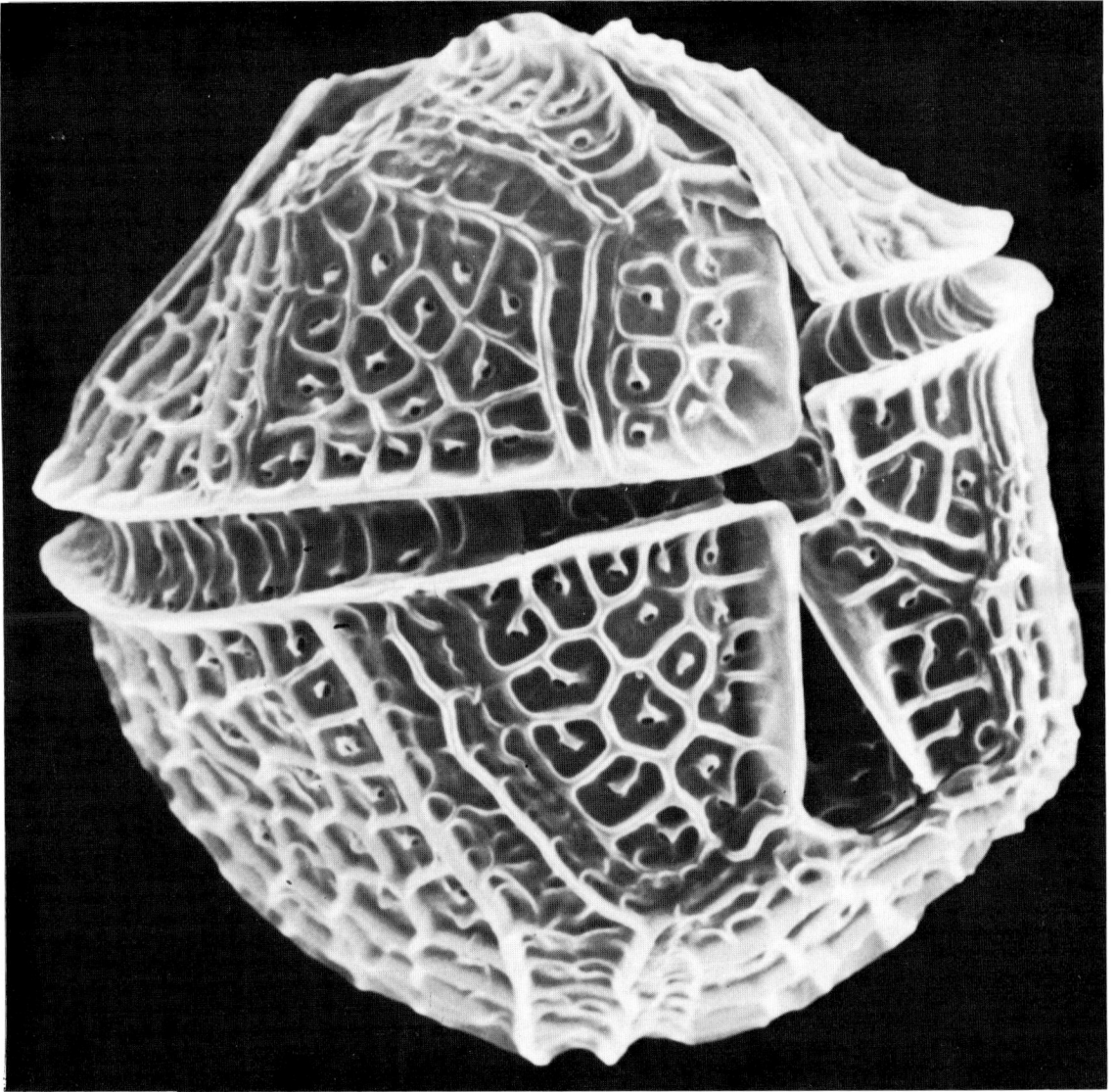

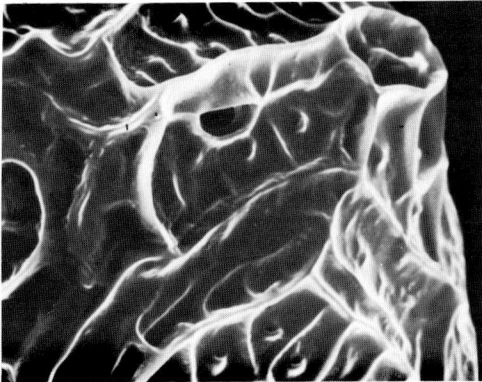

Protoceratium reticulatum (Claparéde & Lachmann) Butschli
(also known as *Gonyaulax grindleyi* Reinecke)
This beautifully ornamented globular dinoflagellate has been the cause of red water off the coast of South Africa and at times can be common in parts of the North Sea. It has an ovoid spiny cyst known as *Operculodinium centrocarpum*. North Sea. 47 µm 1, 47 µm w.

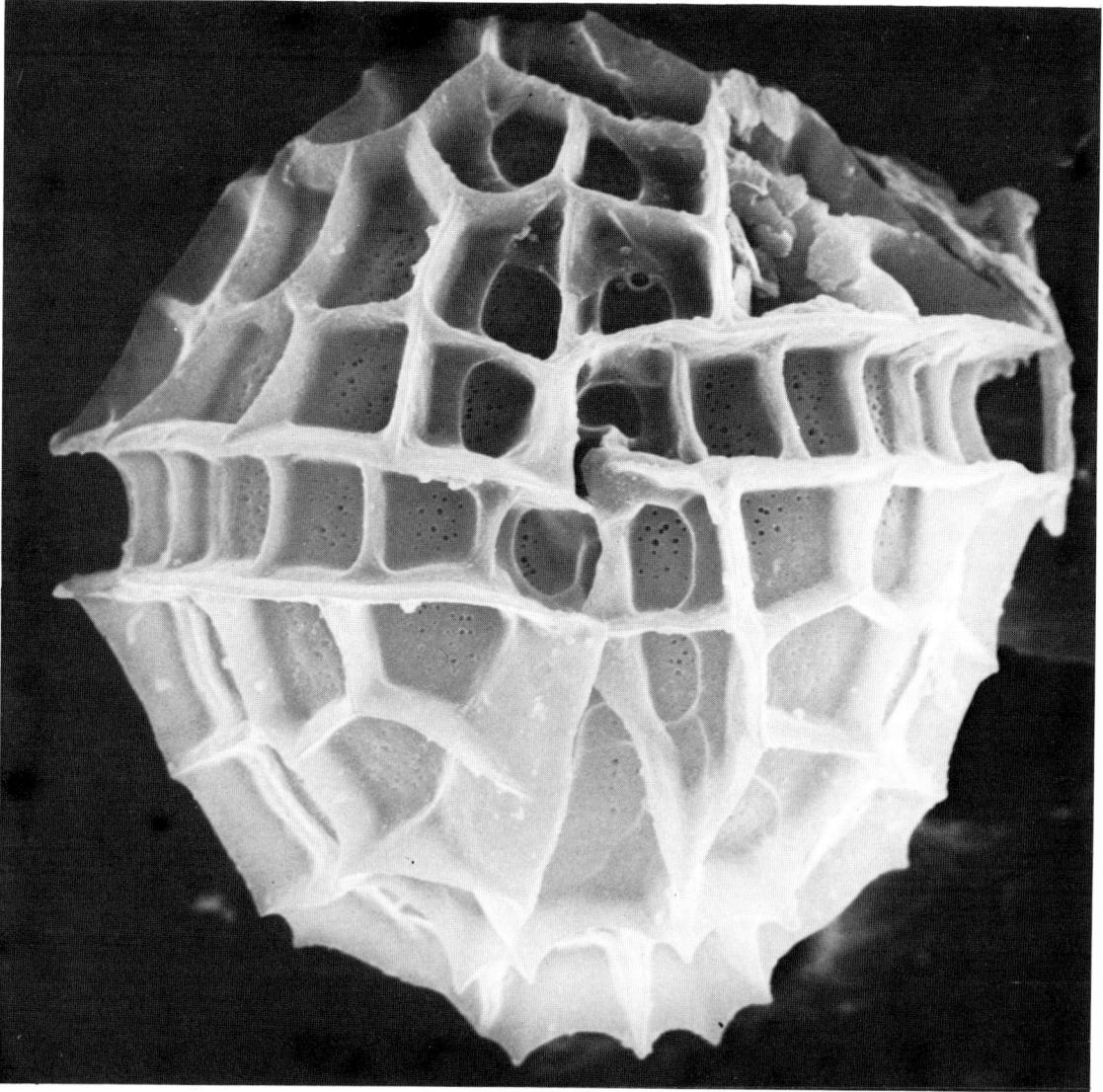

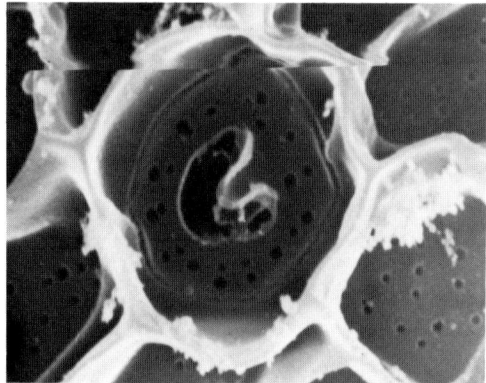

Protoceratium spinulosum (Murray &
Whitting) Schiller
This is a planktonic species of warm
waters. The theca is so heavily ornamented that it is practically impossible to
discern the arrangement of the plates.
The girdle is also unusually strongly
ridged. The small picture shows the apical pore.
E. Atlantic. 35 μm l, 30 μm w.

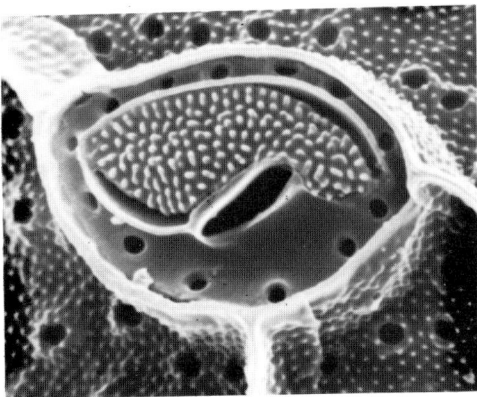

Pyrodinium bahamense Plate
An important tropical dinoflagellate
which is often responsible for lumines-
cent seas in the Carribean and East
Indies. It has also been implicated in
some cases of shell fish poisoning (PSP).
The very characteristic theca has wide
ridges at the edges of the plates, very
wide lists and fine ornamentation. The
organism also has an unusual apical pore
(small picture).
New Guinea. 40 μm l, 60 μm w.

Ceratocorys horrida Stein
A striking organism from warm water plankton which lives up to its name! This is a
gonyaulacoid dinoflagellate in which the epitheca is reduced to a flattened plate but the
hypotheca is extended into a variable number of long ridged spines.
Tropical E. Atlantic. 100 μm 1, 60 μm w.

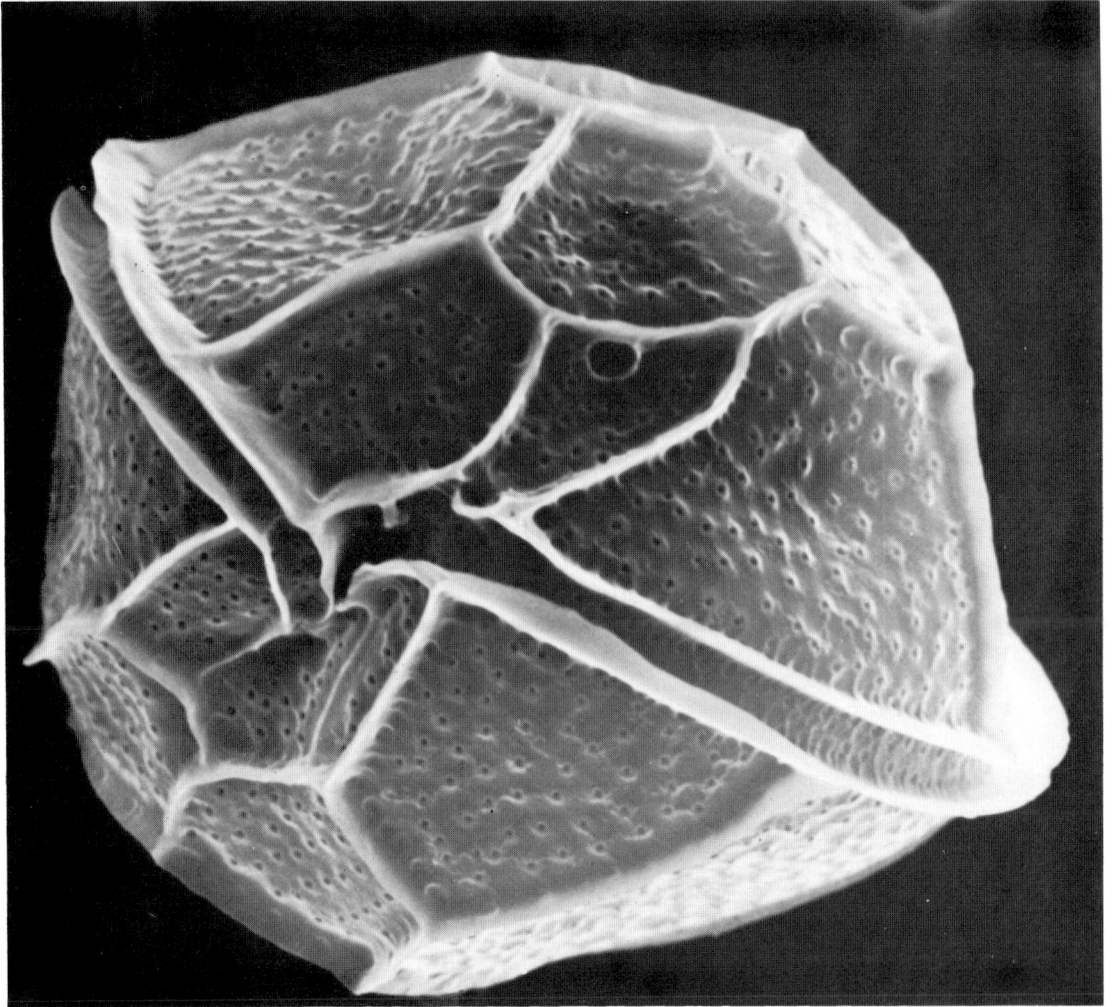

Triadinium polyedricum (Pouchet) Dodge
(also placed in the genera *Gonyodoma* and *Heteraulacus*)
A ventral view of this distinctively angular oceanic dinoflagellate. Note the ridges which
follow the plate sutures and the ventral pore in the plate above the sulcus. Often found in
the Gulf Stream to the west of the British Isles.
Plate formula: Po, 3', 7'', c, s, 5''', 1p, 1''''.
N.E. Atlantic. 60 μm l, 70 μm w.

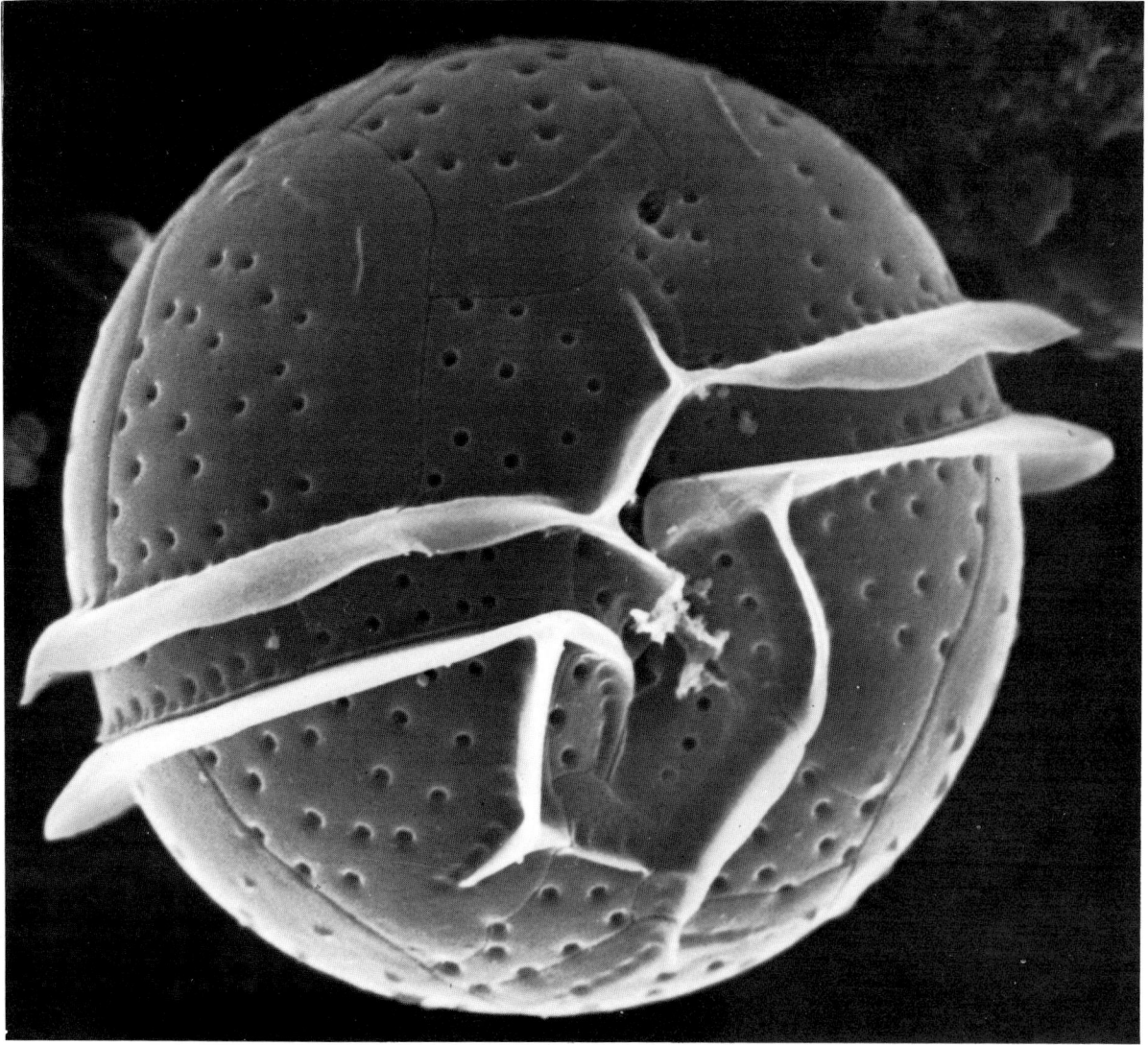

Triadinium sphaericum (Murray & Whitting) Dodge
By way of contrast with *T. polyedricum* the theca here is smooth and the cell distinctly rounded. There are no projections apart from the girdle and sulcal lists. This is a species of warm oceanic waters.
E. Atlantic. 35 μm d.

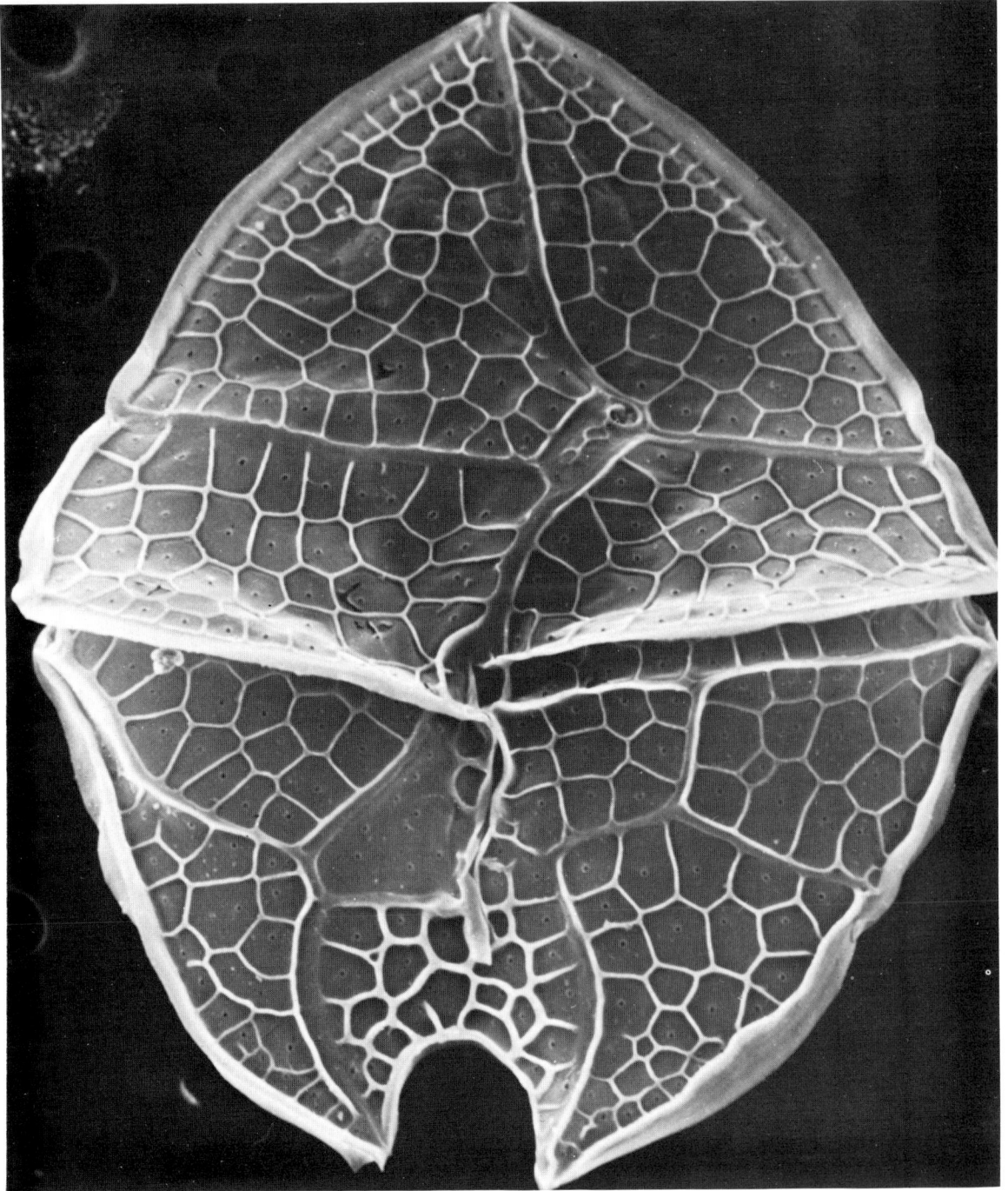

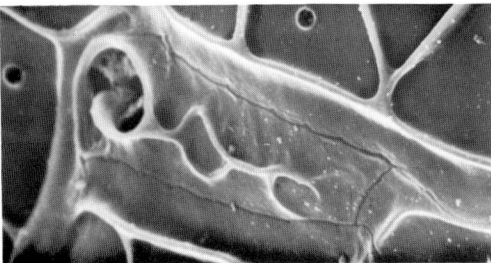

Heterodinium whittingae Kofoid
A remarkable, large, leaf-like dinoflagellate from warmer seas. The tabulation is unusual with well marked reticulations, each enclosing a pore, and the girdle is incomplete on the right side. Note the characteristic ventral pore (small picture).
E. Atlantic. 140 μm l, 115 μm w.

92

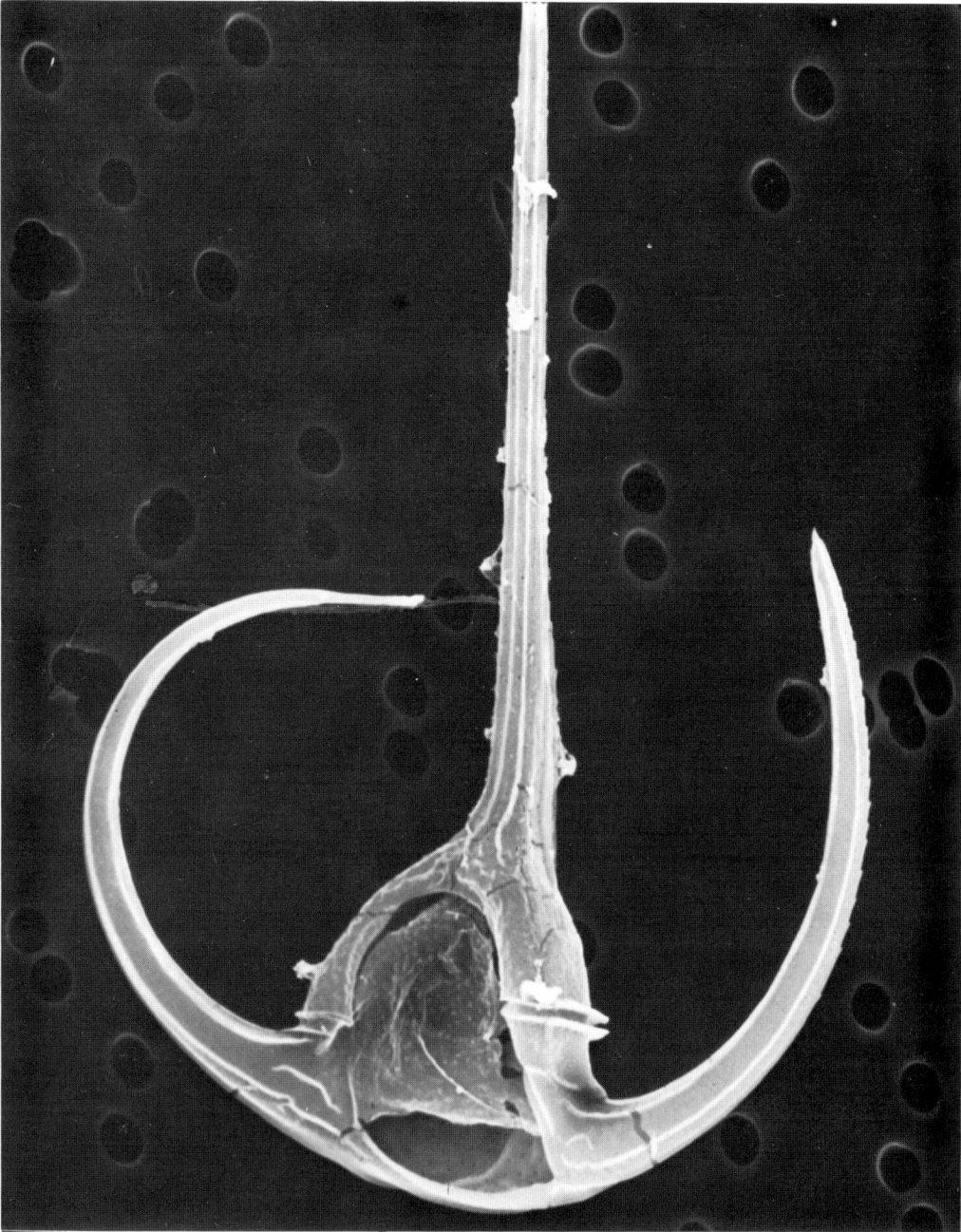

Ceratium arietinum Cleve

This slender, anchor-shaped *Ceratium* is distinguished by having one of its antapical horns bent around to almost touch the apical horn. An oceanic warm water species. Plate formula in this genus: Po, 4', 5'', 4c, s, 5''', 2''''.

E. Atlantic. 270 μm l, 140 μm w.

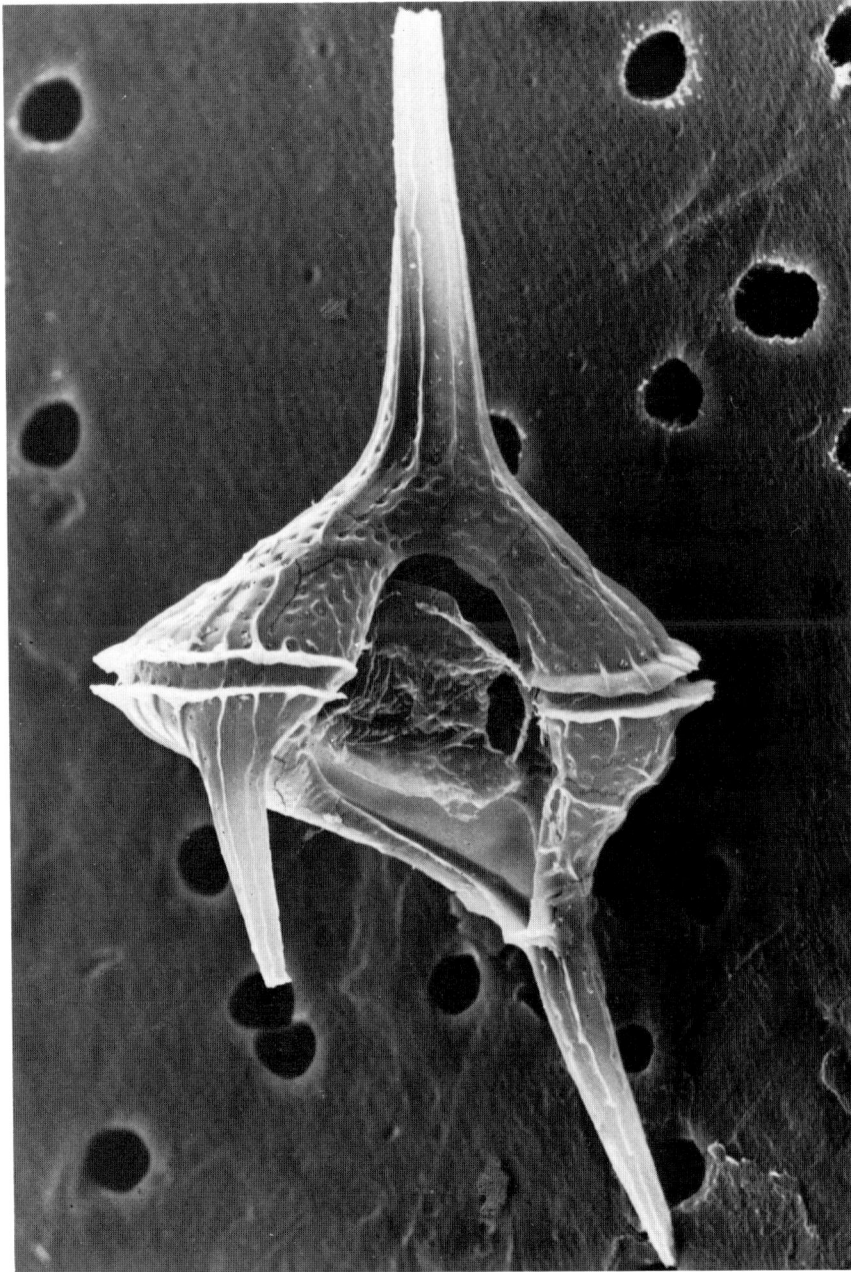

Ceratium candelabrum (Ehrenberg) Stein
A stubby species with two posteriorly directed antapical horns and a short apical horn. As is common in this genus the ventral area of the cell is covered by delicate plates which collapse when the cell is dried. An oceanic organism of warmer waters.
E. Atlantic. 140 μm l, 70 μm w.

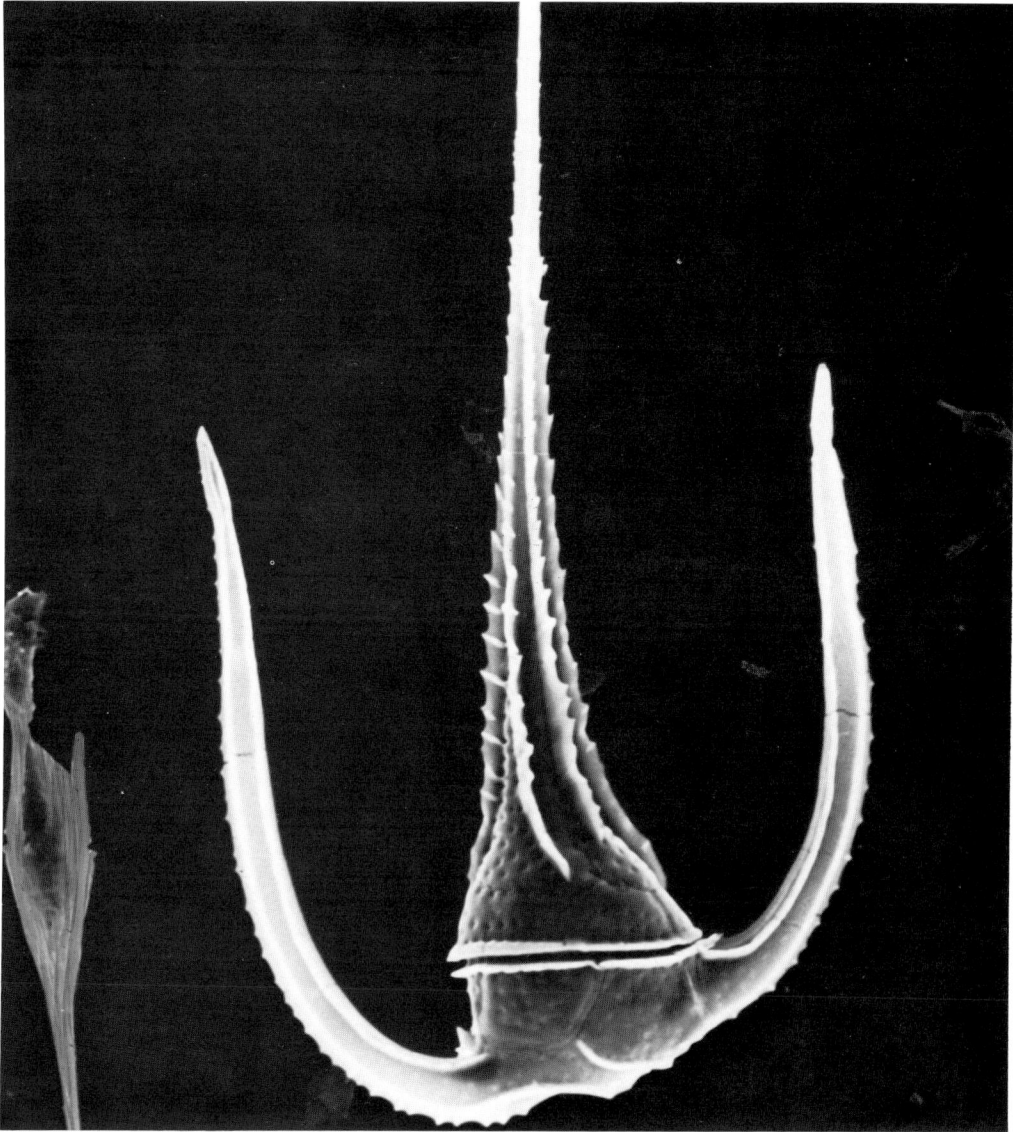

Ceratium compressum Gran

A dorsal view of this oceanic species. It is distinguished from the other anchor-shaped ceratia by the ridged apical horn and the tendency of the pointed antapical horns to become inflated.

N. Atlantic. 250 μm l, 150 μm w.

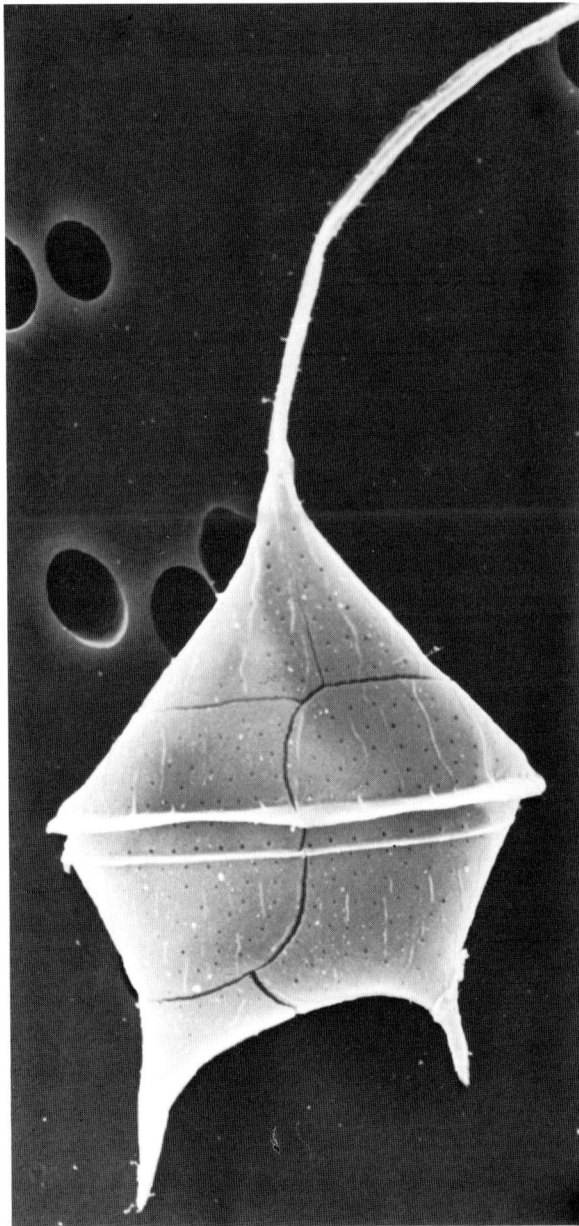

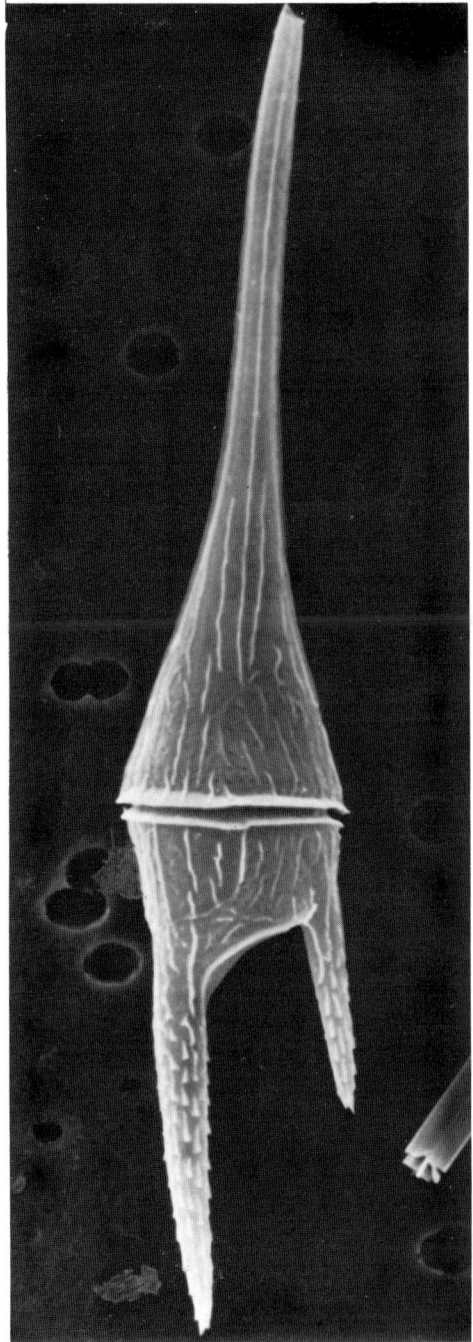

Ceratium furca (Ehrenberg) Claparédè & Lachmann (l);
C. pentagonum Gourret (r).
Ventral views of these species both of which have more or less parallel antapical horns. *C. furca* is common around the British Isles and *C. pentagonum* is normally found in warmer waters.
C. furca: North Sea 230 µm l, 40 µm w.
C. pentagonum: E. Atlantic 125 µm l, 50 µm w.

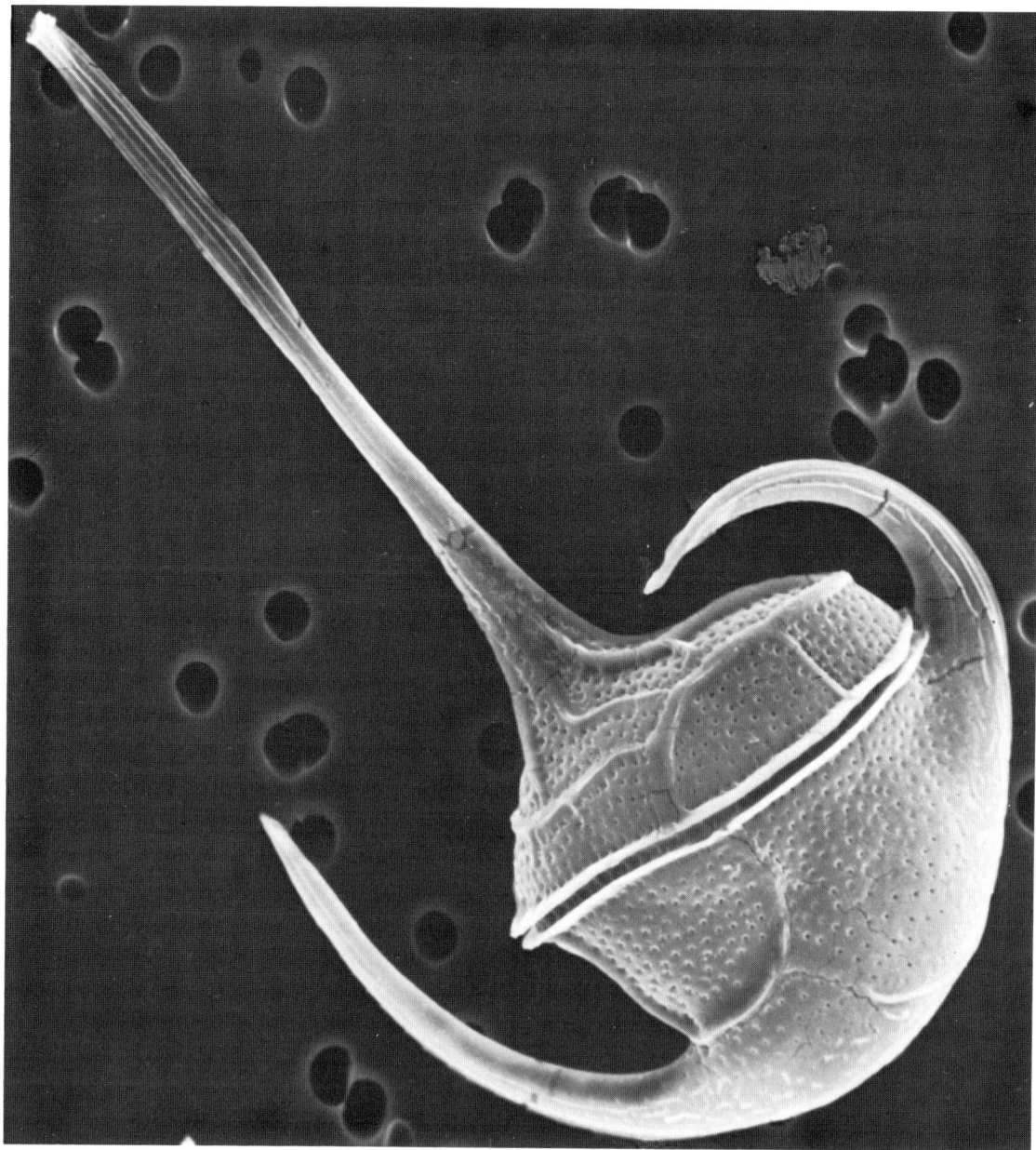

Ceratium gibberum Gourret
A warm-water species with a rather bulbous body on which the anterior horn is placed asymmetrically. The right antapical horn is bent around so that it almost touches the epitheca. This dorsal view clearly shows how the antapical horns are extensions of hypothecal plates.
Tropical E. Atlantic. 230 μm l, 125 μm w.

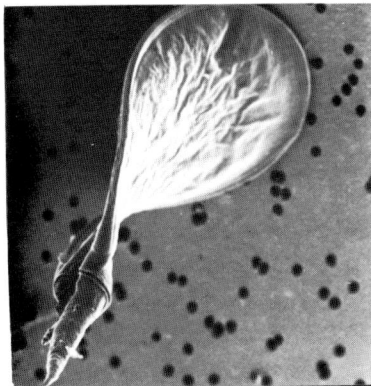

Ceratium gravidum Gourret
An extremely large tropical dinoflagellate
in which the epitheca is both inflated and
dorso-ventrally flattened to give the bag-
shaped structure shown. Surprisingly the
girdle and hypotheca are of normal size
for the genus. The small picture is a side
view and shows how thin the cell is.
Tropical E. Atlantic. 340 μm l, 170 μm w.

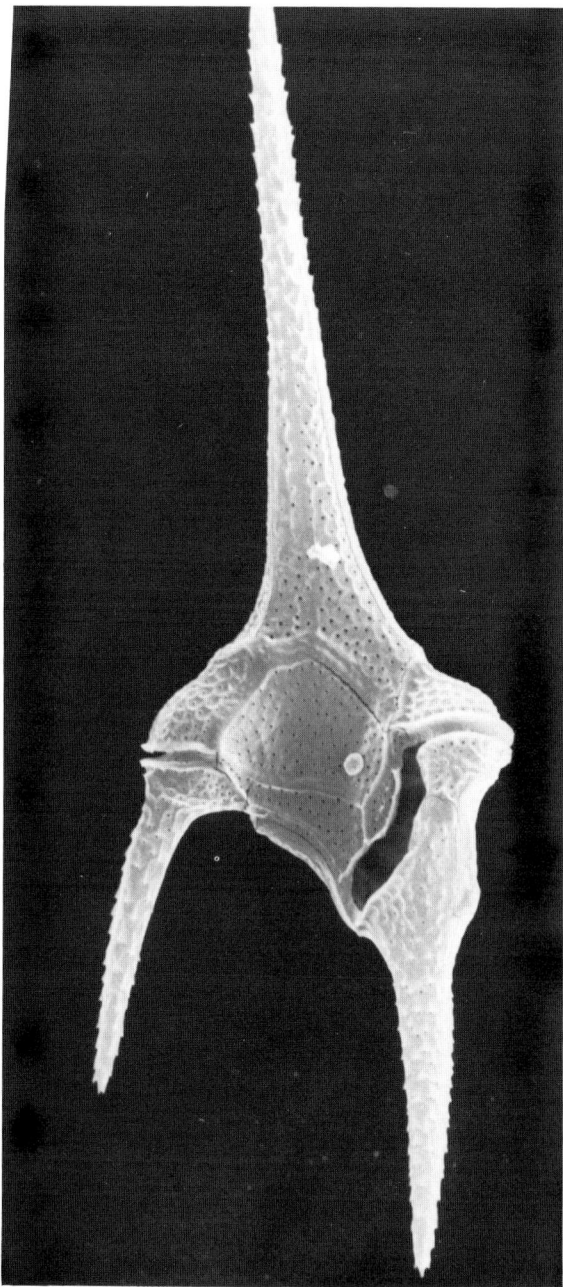
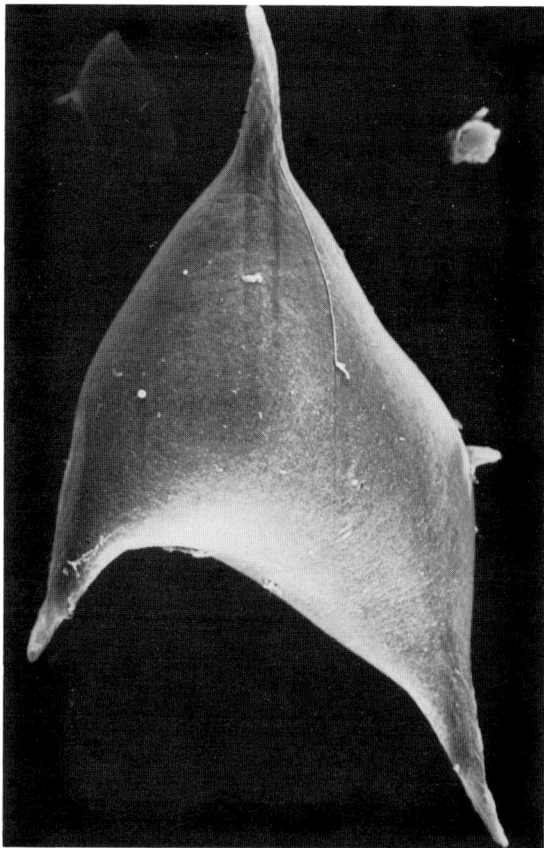

Ceratium hirundinella O. F. Müller
A fairly common freshwater dinoflagellate which can be so numerous as to colour lake water brown. It is a rather variable species and this shows a narrow form. The organism overwinters by forming a cyst (smaller picture) which sinks to the bottom of the lake and may germinate the following spring to give rise to a new motile cell.
Esthwaite, Cumbria. (F.W.) 200 μm l, 60 μm w.

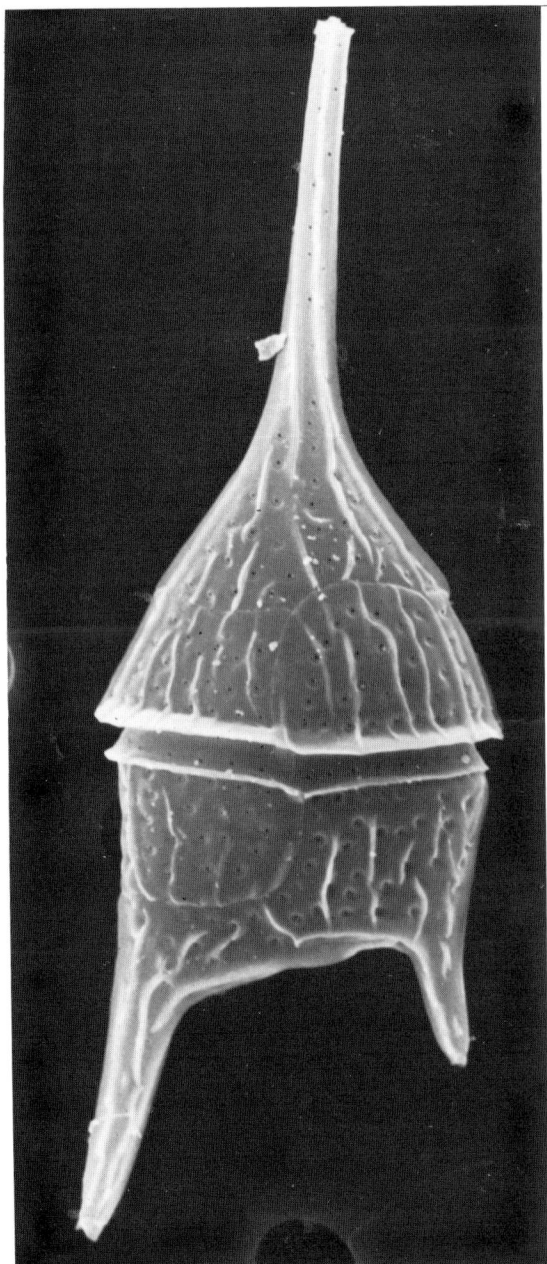

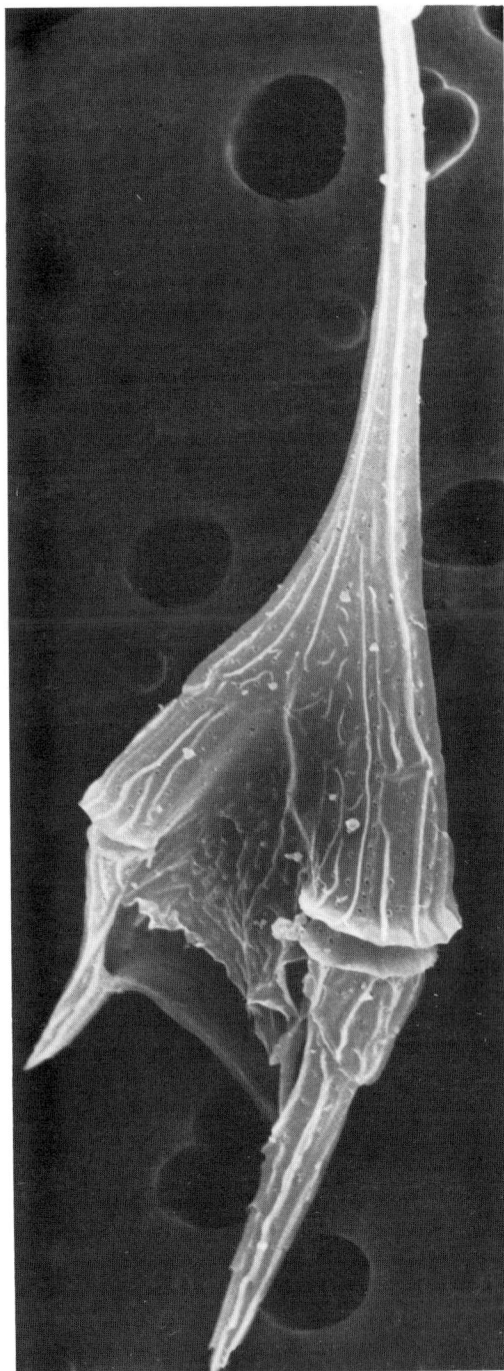

Ceratium lineatum (Ehrenberg) Cleve
Dorsal (l) and ventral (r) views of this delicate species which is one of the smallest in the genus. In the dorsal view the antapical horns can be seen to be slightly angled (compare with *C. furca*). Often quite common in the seas around the British Isles, particularly to the N.W.
North Sea. 95 μm l, 30 μm w.

100

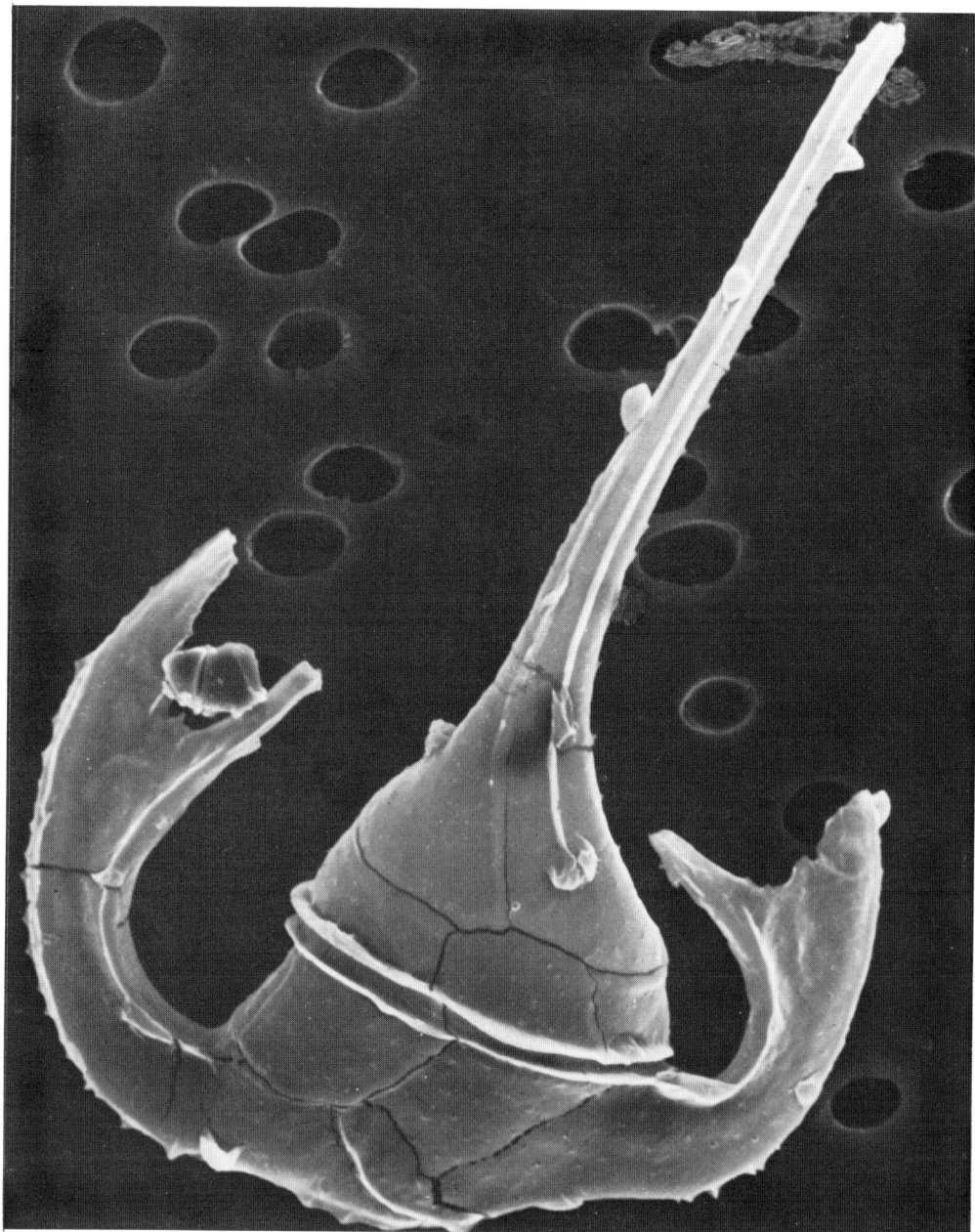

Ceratium platycorne Daday
A species in which the antapical horns are grossly inflated whereas the apical horn looks similar to that of many other species. This dorsal view shows rather clearly the sutures between the plates. A warm-water form.
Tropical E. Atlantic. 100 μm l, 160 μm w.

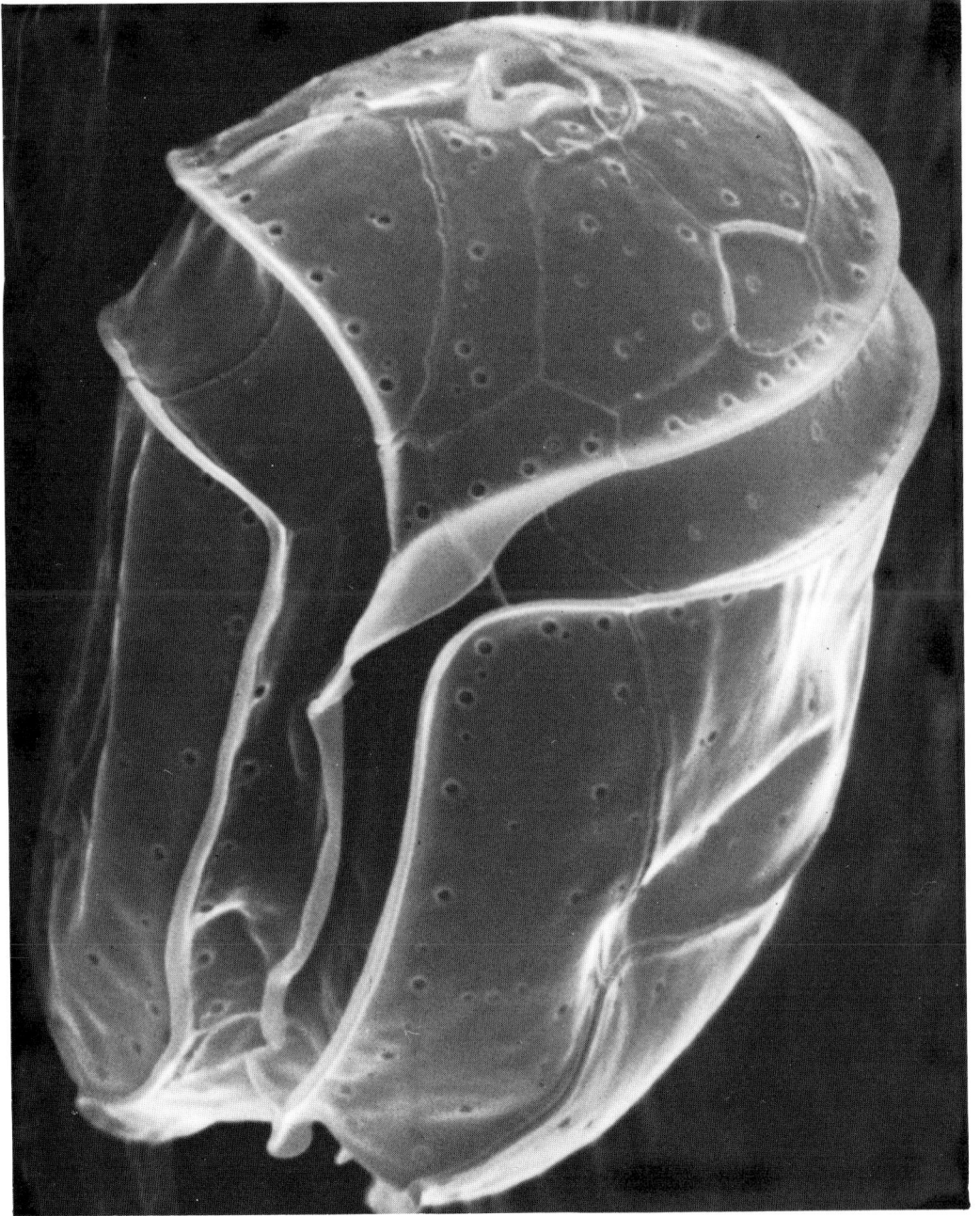

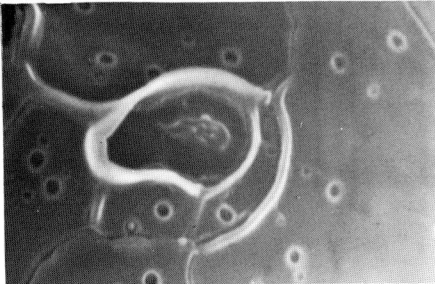

Amphidiniopsis kofoidii Woloszynska
A sand-dwelling dinoflagellate which is
not so flattened and has smoother thecal
plates than the other members of this
genus. An unusual feature is the way in
which the epitheca is extended into the
sulcus where it forms a central wing.
The small picture shows the apical pore.
N. Brittany coast. 20 μm l, 14 μm w.

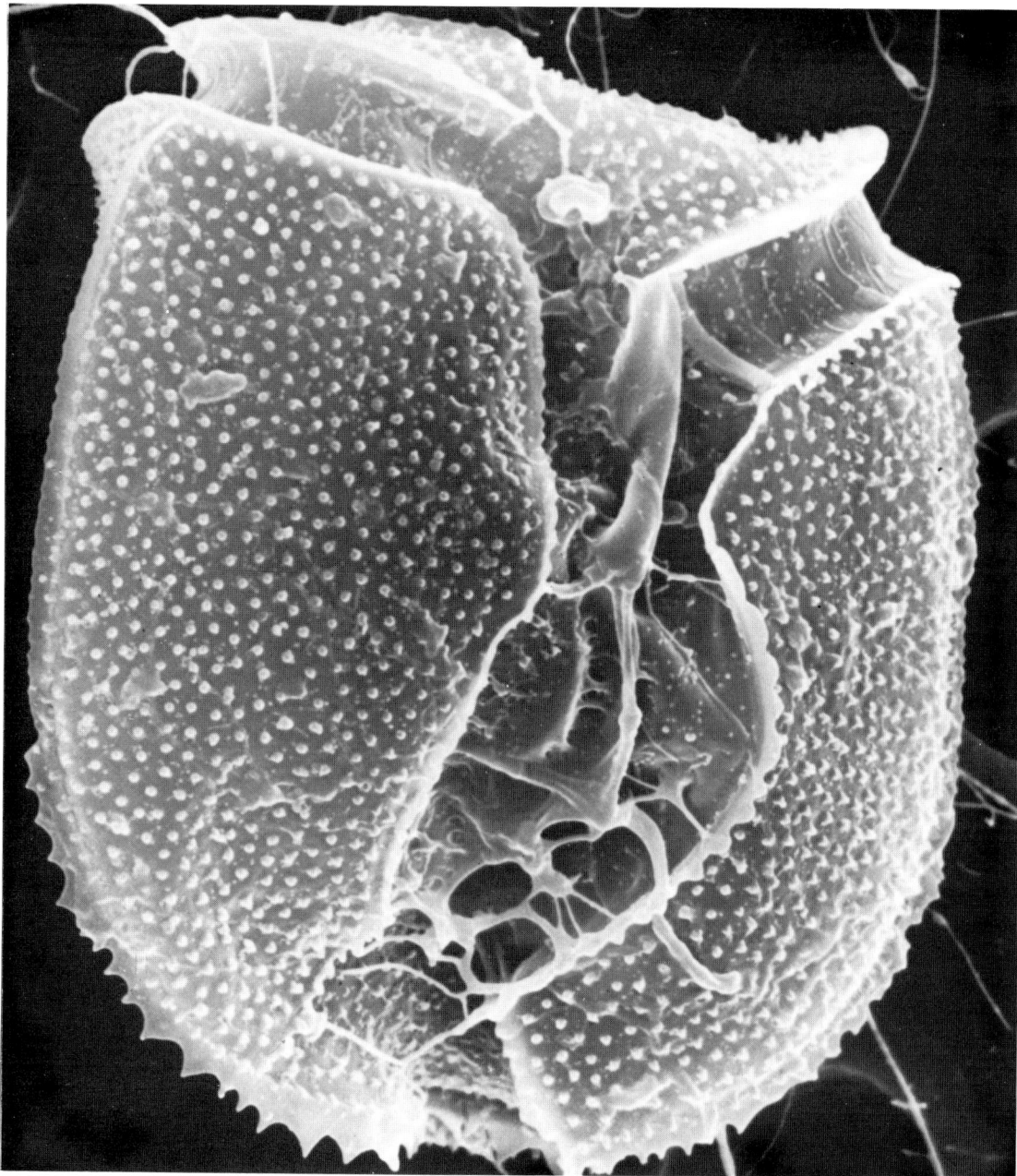

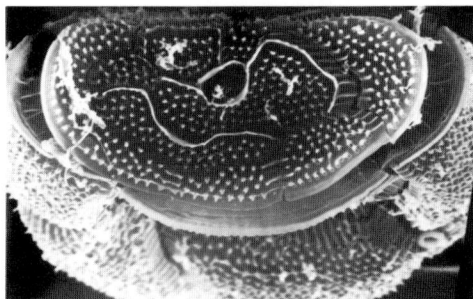

Amphidiniopsis hirsutum (Balech)
Dodge
A dorso-ventrally flattened dinoflagellate
which is found on sandy beaches. The
epitheca (small picture) is much reduced
and flattened and the hypotheca with its
pimpled plates is cut by the sulcus.
Roscoff, Brittany. 30 μm l, 25 μm w.

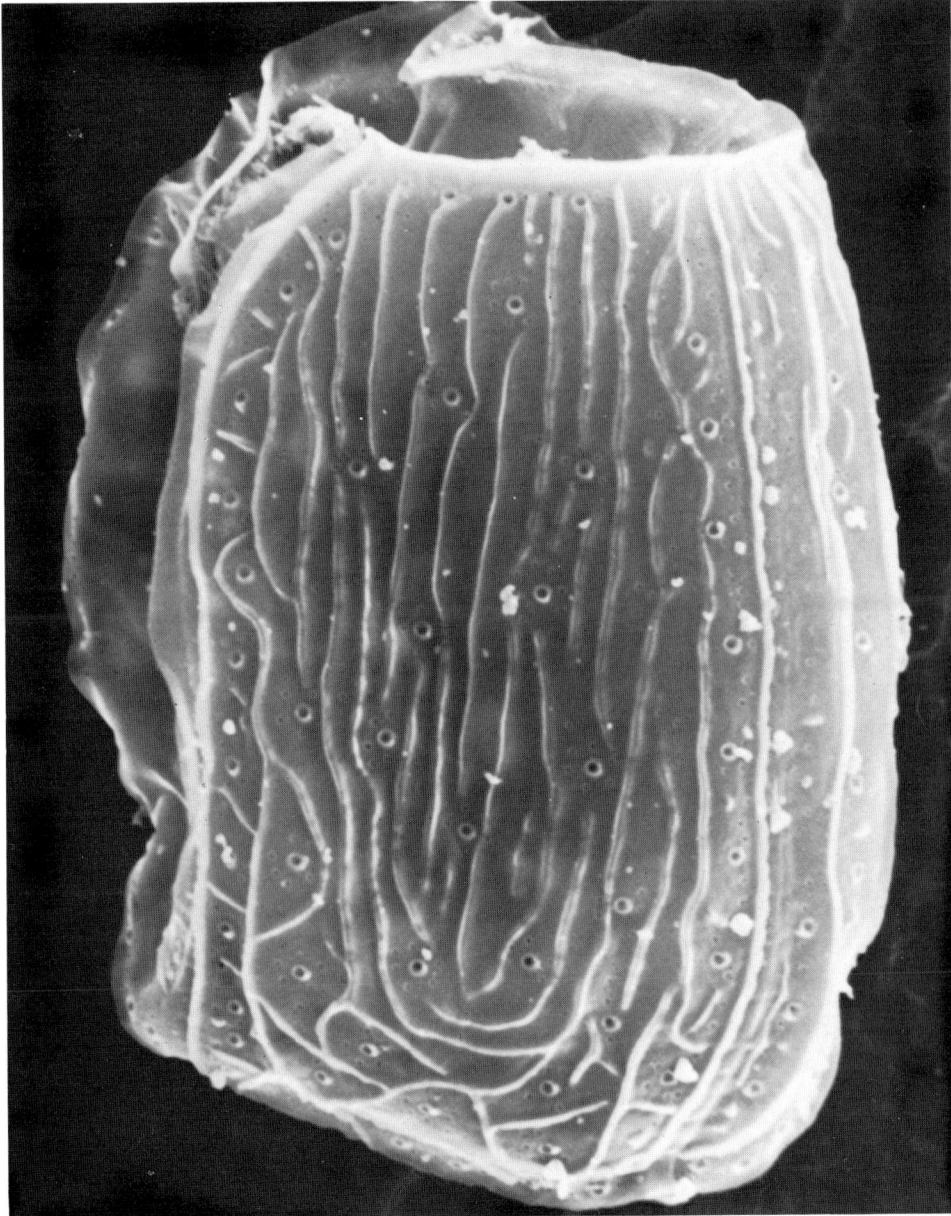

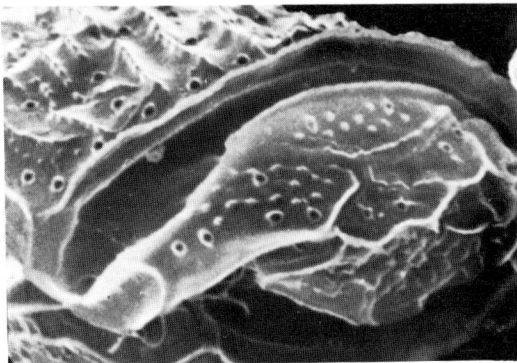

Planodinium striatum Saunders &
Dodge
A newly described bilaterally flattened
sand-dwelling dinoflagellate with a very
much reduced epitheca and large ridged
hypothecal plates. The sulcus is to the
left of the picture. The small photo
shows the epitheca.
Millport beach, Scotland. 24 μm l, 16 μm w.

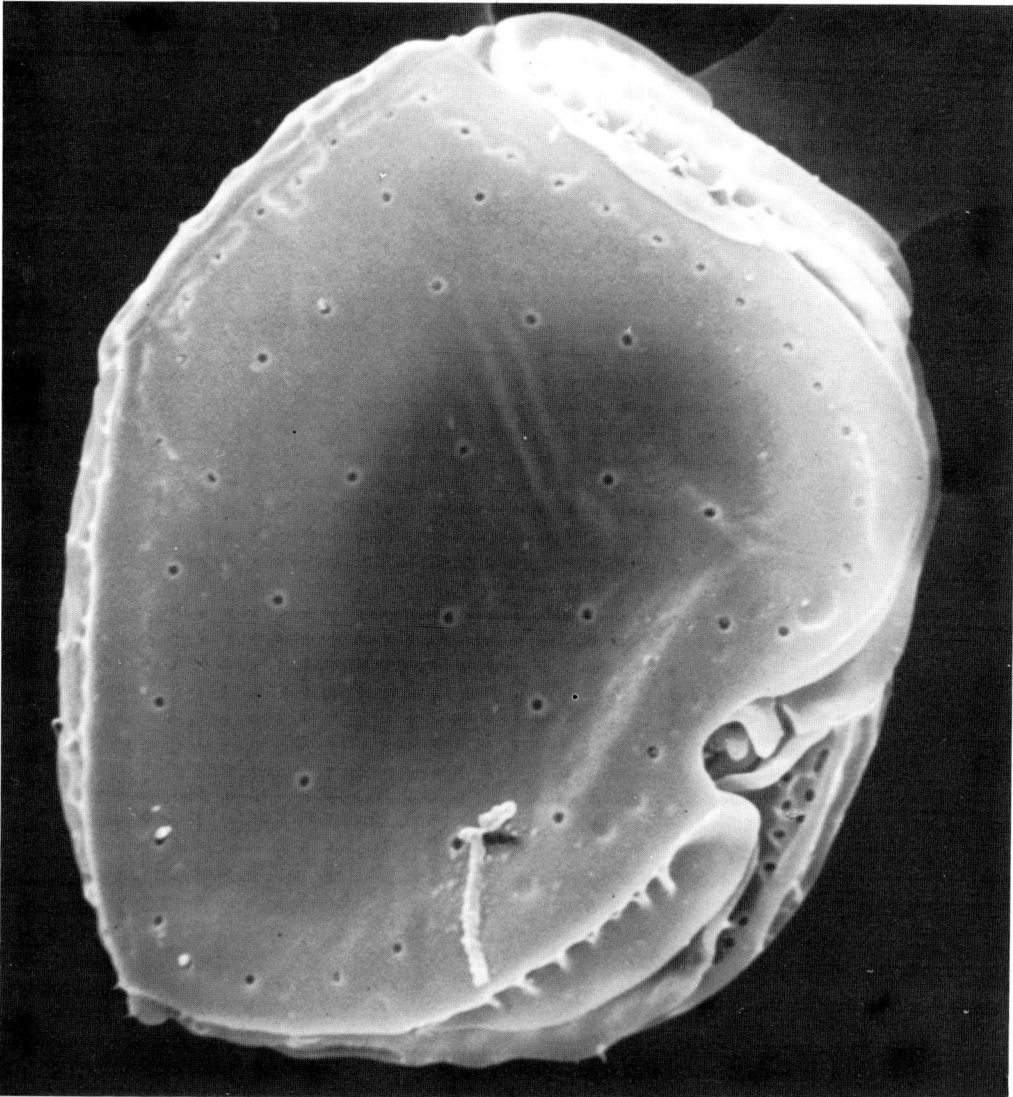

Thecadinium kofoidii Kofoid &
Skogsberg
A bilaterally flattened sand dinoflagellate
here seen from its right side with the
pore of the longitudinal flagellum just
visible. The epitheca (small picture) is
very much reduced and the hypotheca
consists of just four plates. Some workers
have incorrectly placed this organism in
the Dinophysiales.
Millport, Scotland. 30 μm l, 25 μm w.

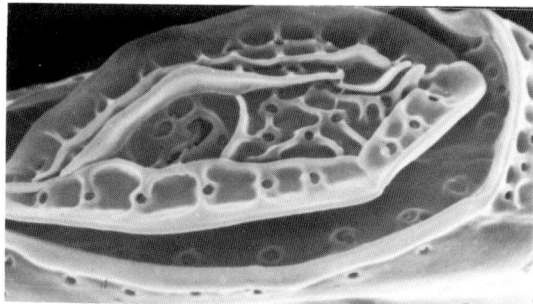

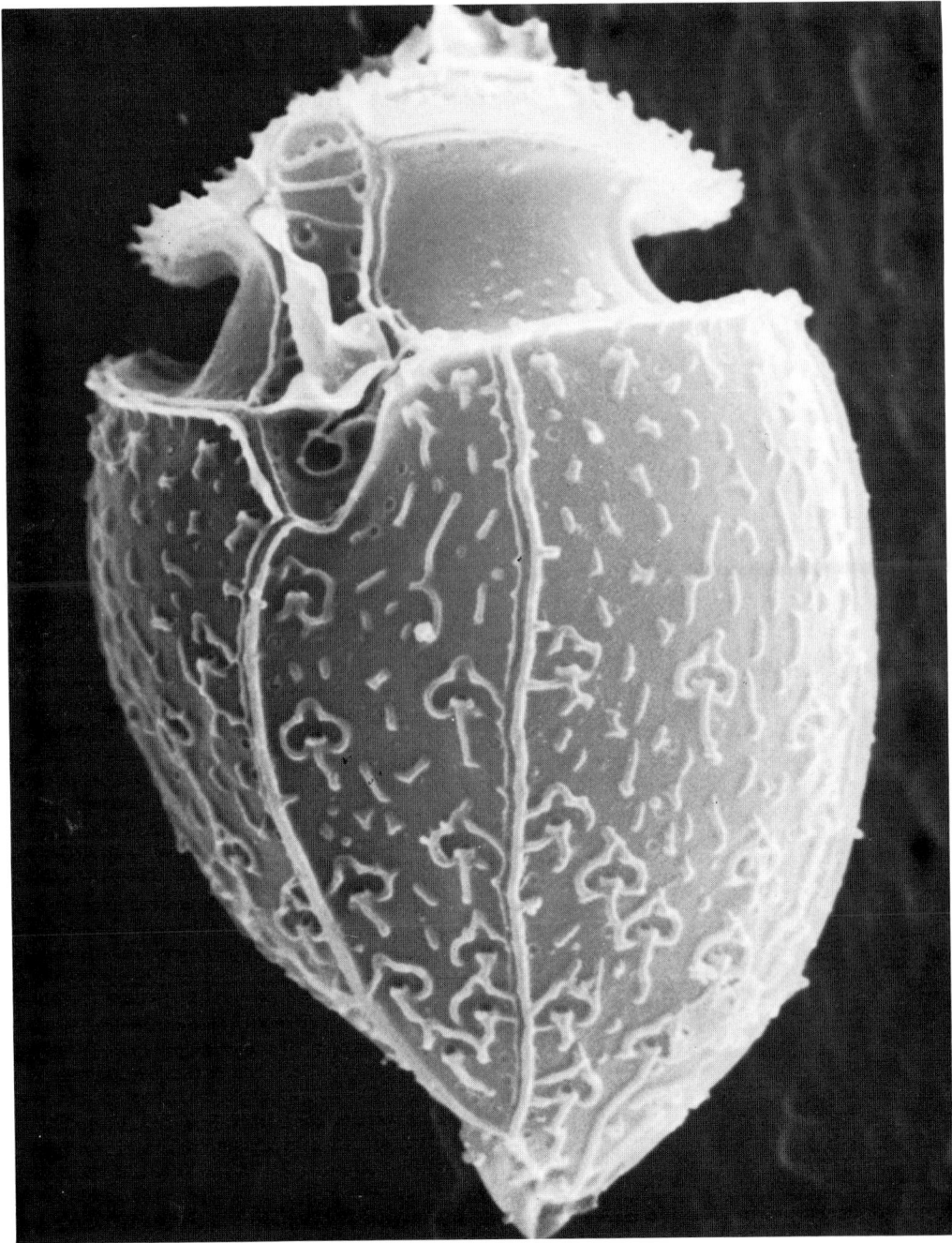

Oxytoxum crassum Schiller

This genus is essentially confined to oceanic water and the many species vary mainly in
their apices and ornamentations. *O. crassum* is quite a simple member with a very well
marked girdle and a delightful ornamentation of the thecal plates.

Plate pattern in the genus: Po, 5', 6'', 5c, 4s, 5''', 1''''.

E. Atlantic 17 μm l, 12 μm w.

106

Oxytoxum constrictum Stein
A distinctive species with its fairly sub-
stantial cap-like epitheca and a hypotheca
which is both ridged and constricted.
Note the ridges within the girdle and the
change in ornamentation either side of
the constriction. The small picture shows
an apical view of the epitheca of *O. reticulatum* (see frontispiece).
E. Atlantic. 45 μm l, 25 μm w.

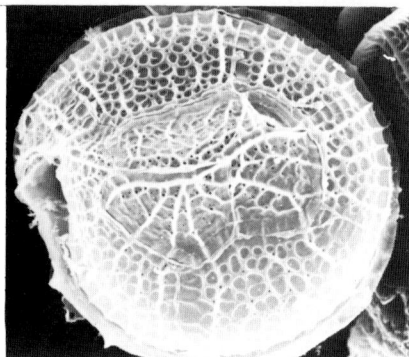

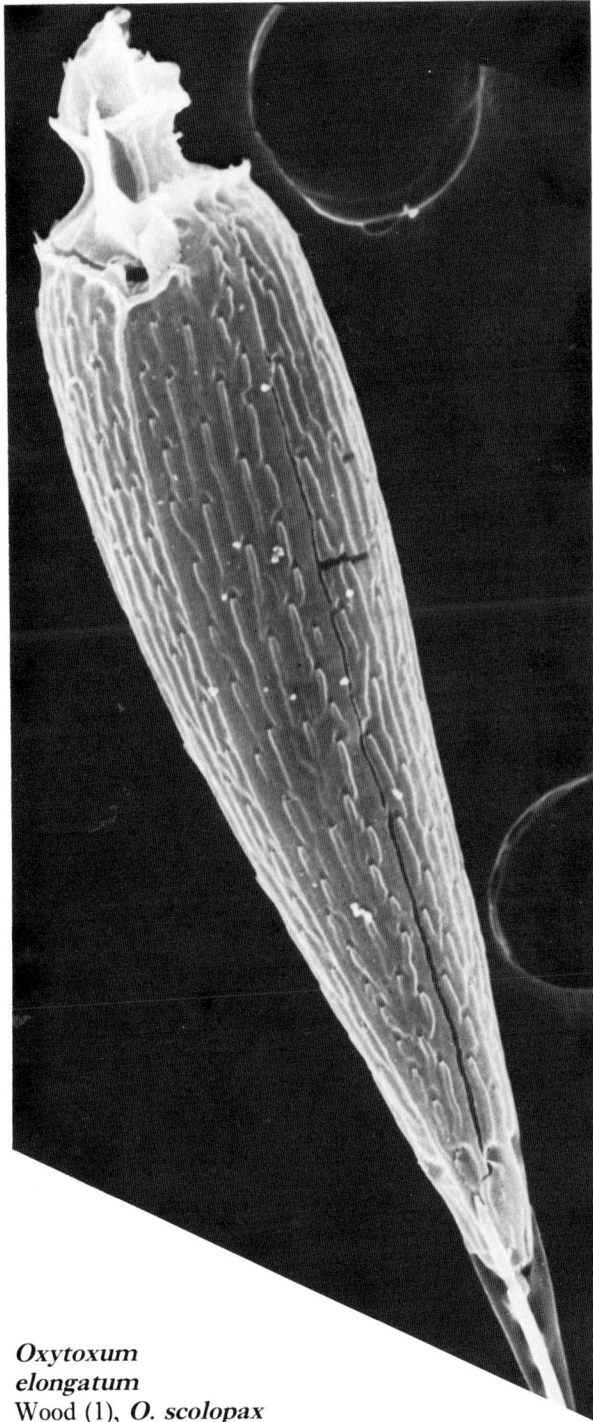

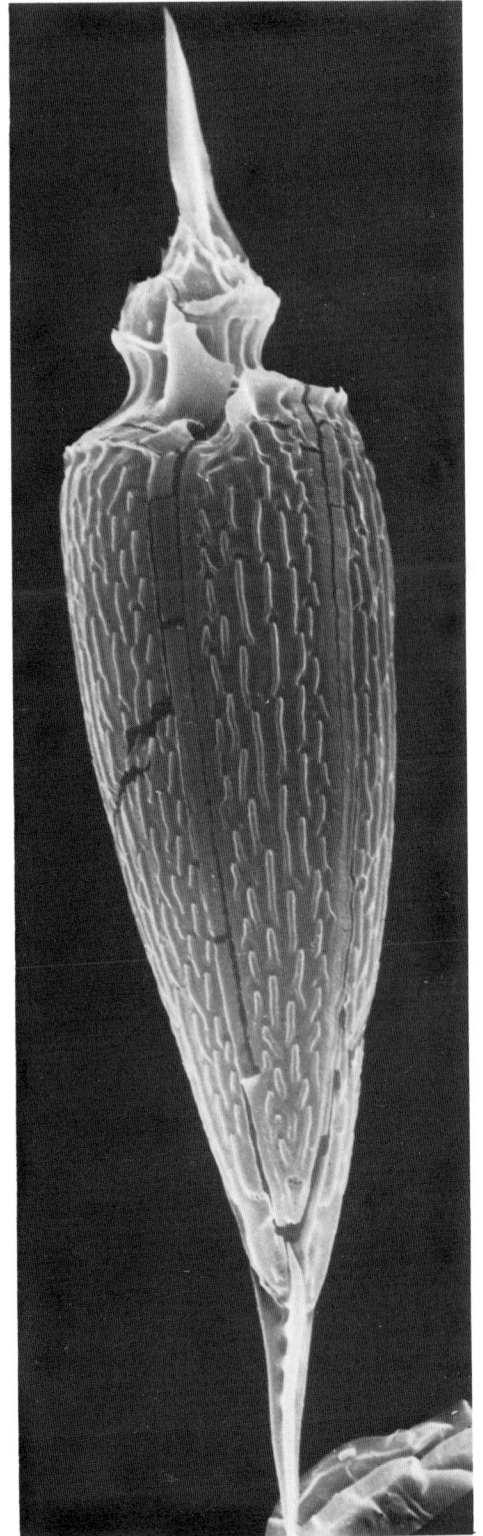

Oxytoxum
elongatum
Wood (1), ***O. scolopax***
Stein (r)
O. elongatum is a long species
with a much reduced epitheca.

E. Atlantic. 65 μm l, 12 μm w.

O. scolopax type species of the genus, is spear shaped with a pointed epitheca.

E. Atlantic. 70 μm l, 15 μm w.

108

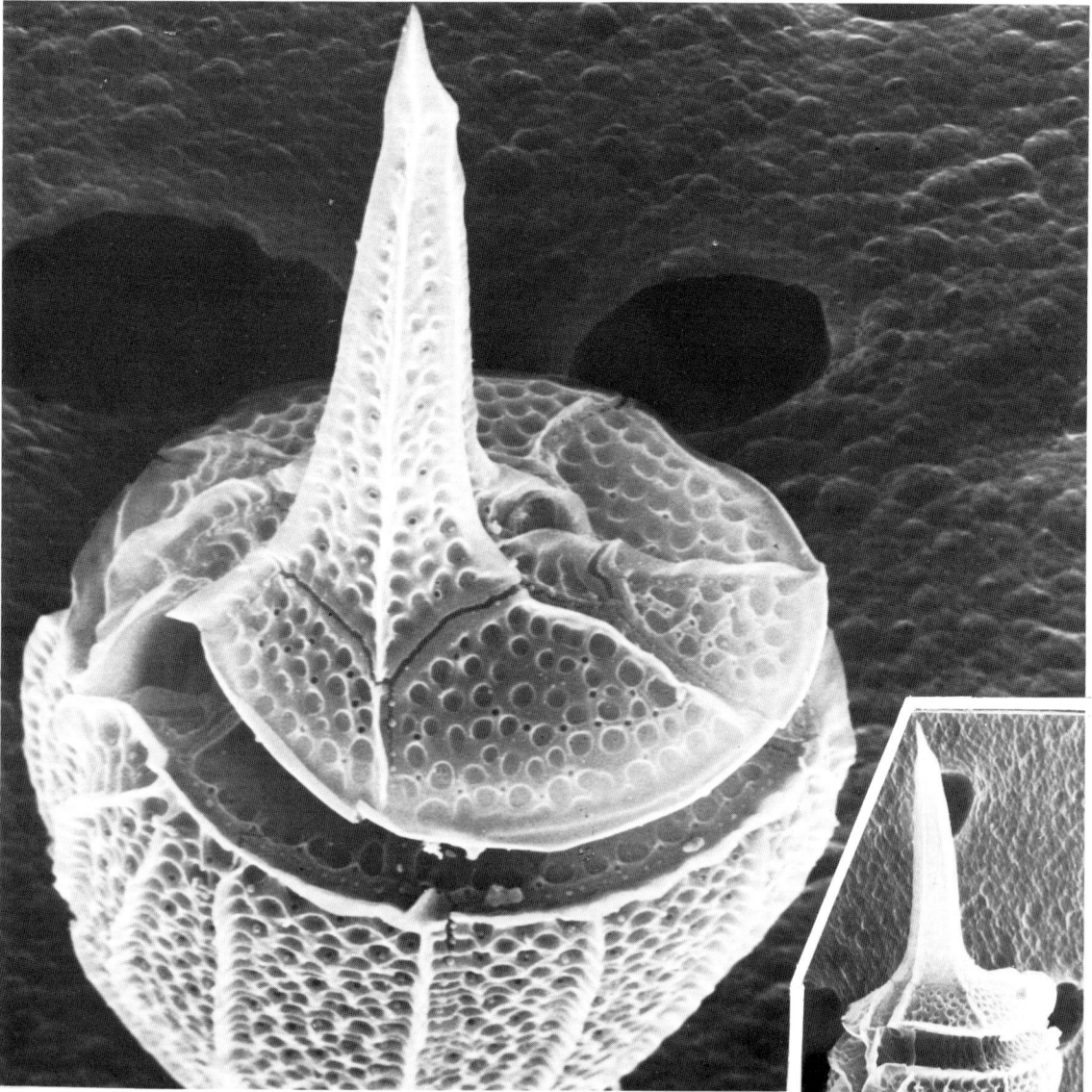

Oxytoxum milneri Murray and Whitting
A large oceanic dinoflagellate with a characteristic spire-like epitheca and gently tapered hypotheca. The plates are ornamented with longitudinal ridges and numerous rounded reticulations.
E. Atlantic. 75 μm l, 25 μm w.

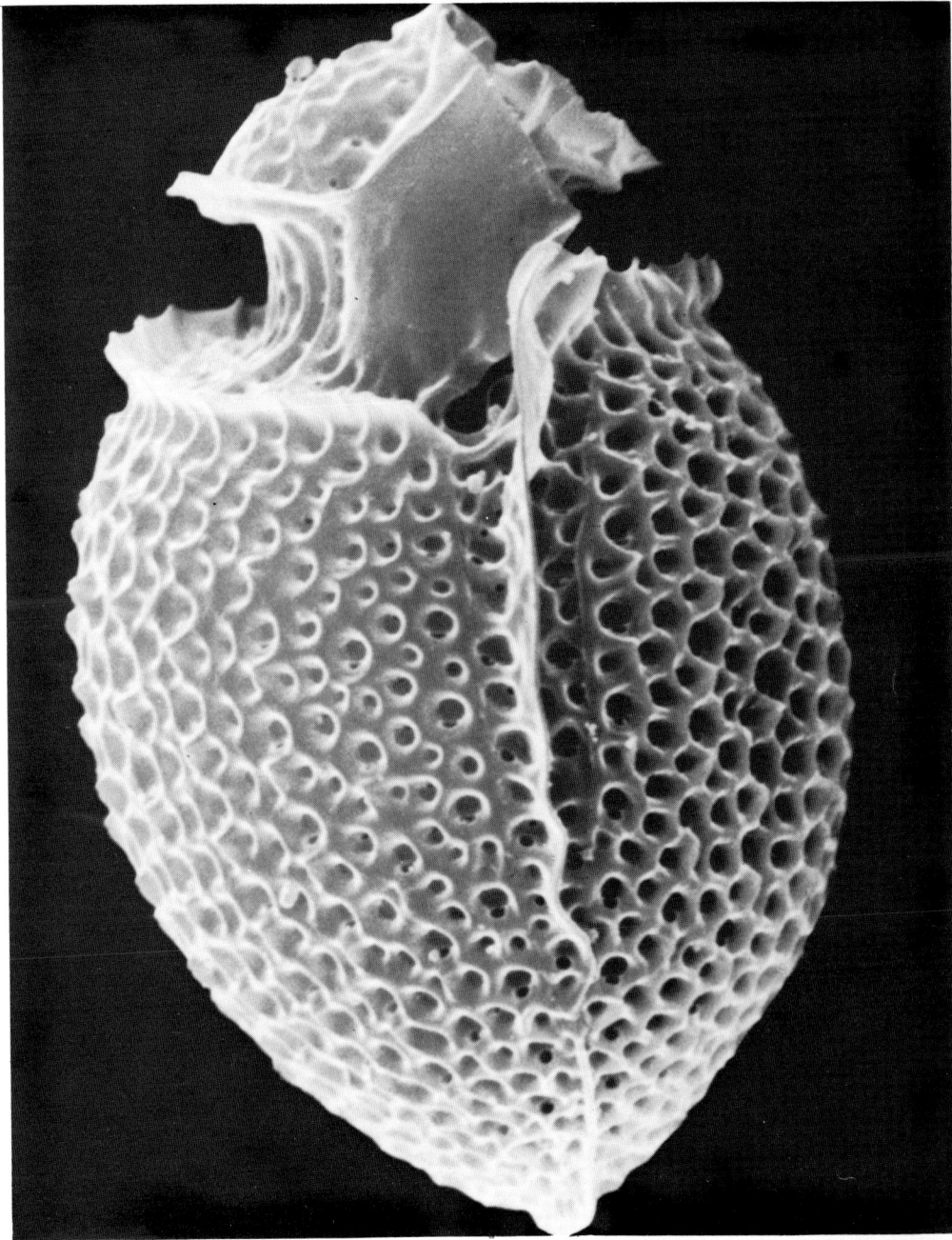

Oxytoxum ovale Schiller
An ovoid cell with small epitheca, very
wide girdle and large sulcal list. The
hypotheca is strongly ornamented with
deep circular reticulations and the antapical plate is reduced to a small ridge.
The small picture shows an antapical
view.
E. Atlantic. 21 µm l, 13 µm w.

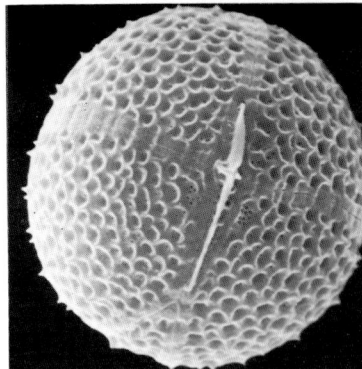

110

Oxytoxum sceptrum (Stein) Schröder
A dorsal view of this short, stout species which has a much reduced epitheca. Note the patterning of the hypotheca which consists of longitudinal ridges and rounded depressions between. E. Atlantic. 35 μm l, 17 μm w.

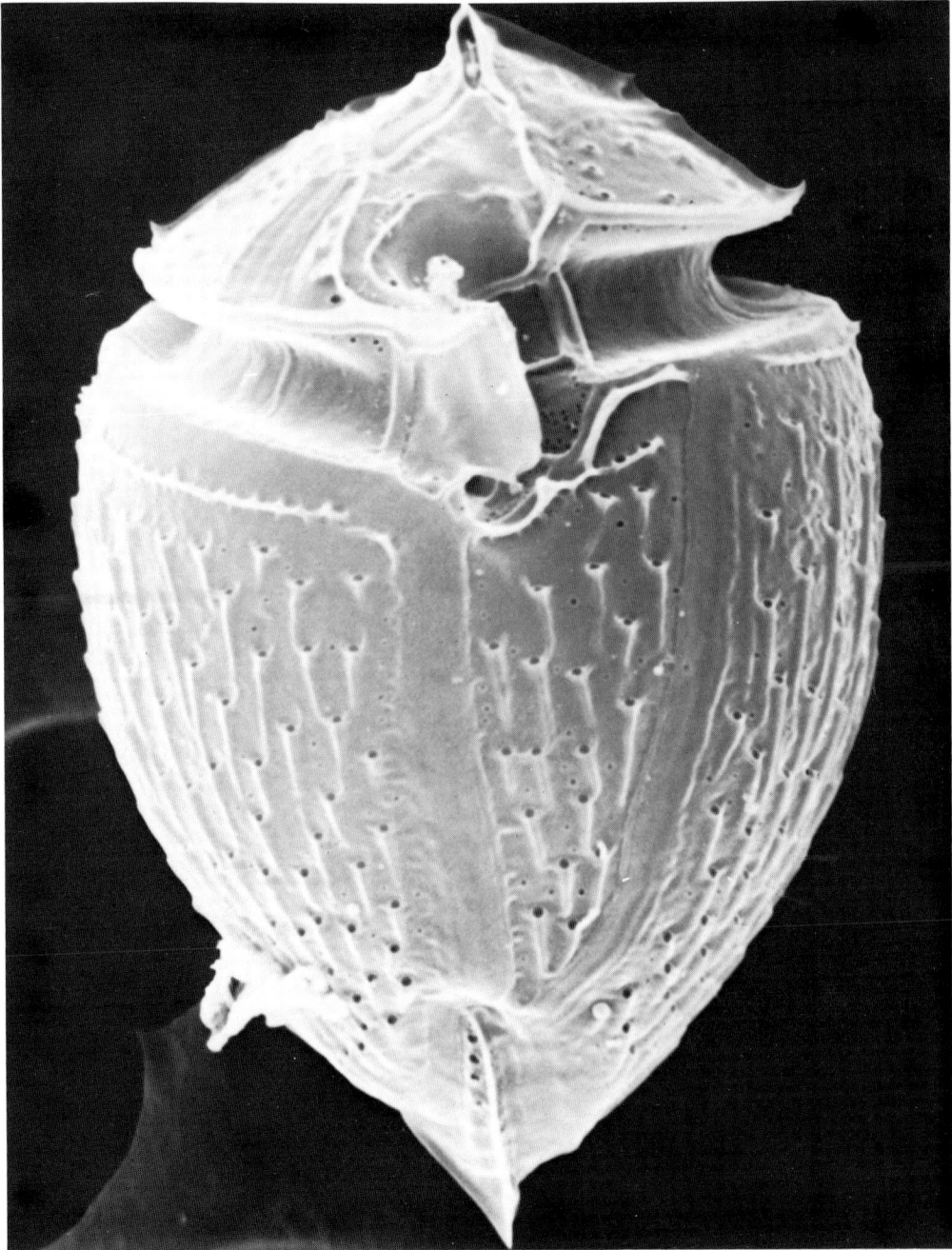

Oxytoxum stropholatum Dodge and Saunders
A top-shaped species with lightly ornamented plates which has recently been discovered in the Atlantic. Note the sulcal wing which covers the flagellar pores and the short longitudinal ridges on the hypotheca which are similar to the markings on *O. scolopax*.
E. Atlantic. 29 μm l, 18 μm w.

112

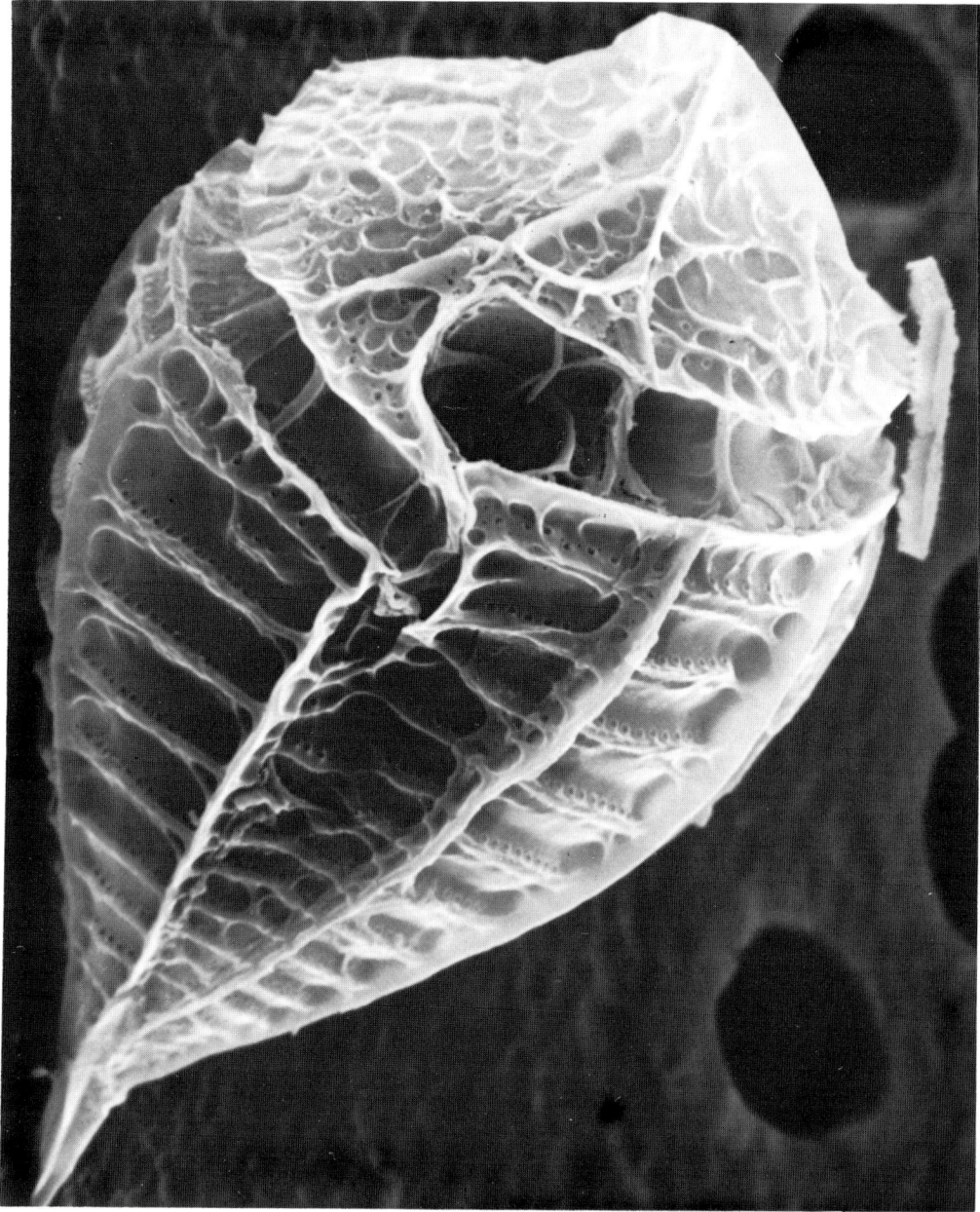

Oxytoxum tesselatum (Stein) Schütt
Here the short cap-like epitheca, offset
girdle and strongly ridged theca charac-
terise the species. Note the longitudinal
and transverse ridges on the hypotheca
which are also shown in the small picture, an antapical view of the hypotheca.
E. Atlantic. 50 μm l, 30 μm w.

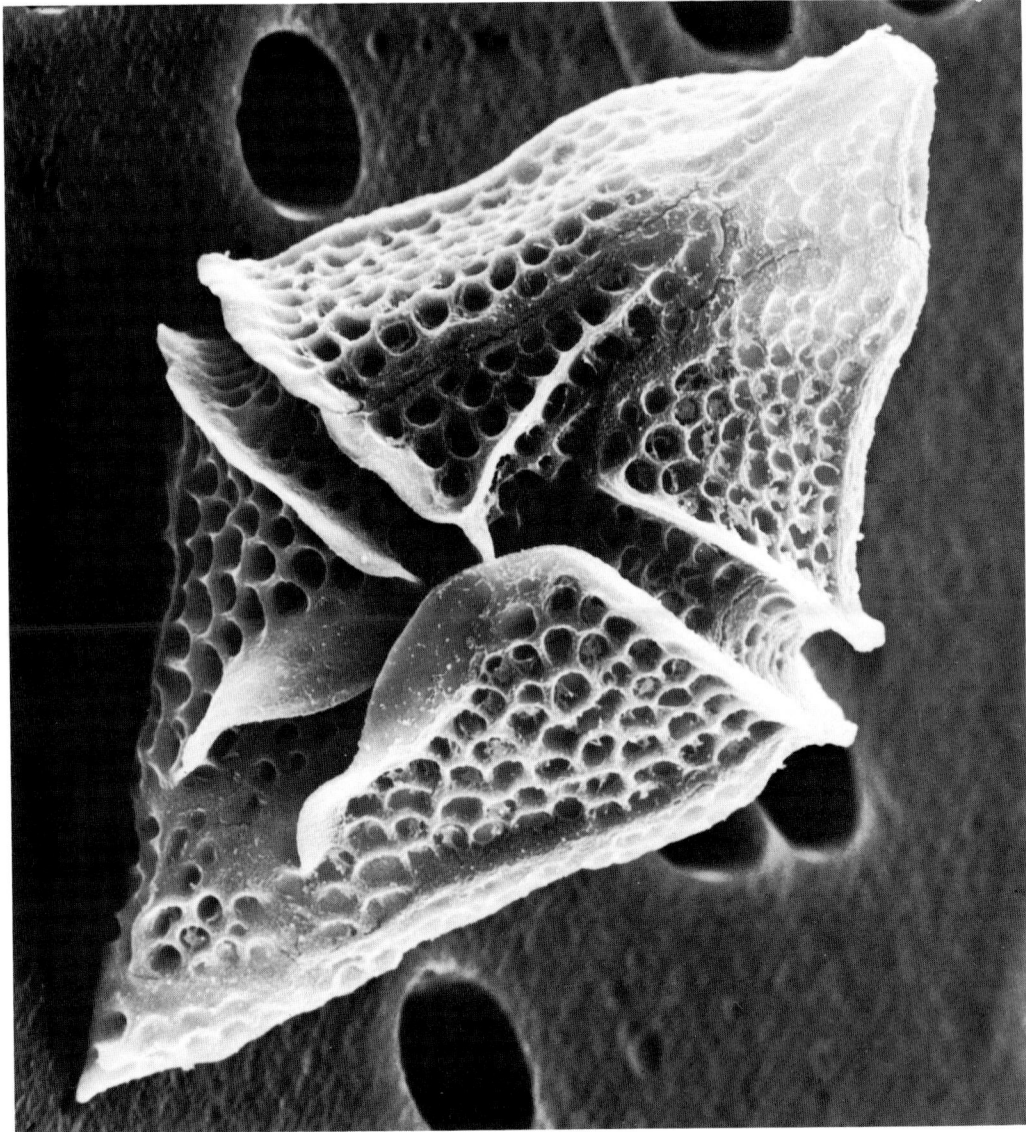

Centrodinium mimeticum (Balech) Taylor
A heavily plated oceanic organism with a slightly offset girdle, which is only known from warm waters. Note the distinctly biconical shape and the lists on both sides of the sulcus as well as above and below the girdle.
E. Atlantic. 60 μm l, 35 μm w.

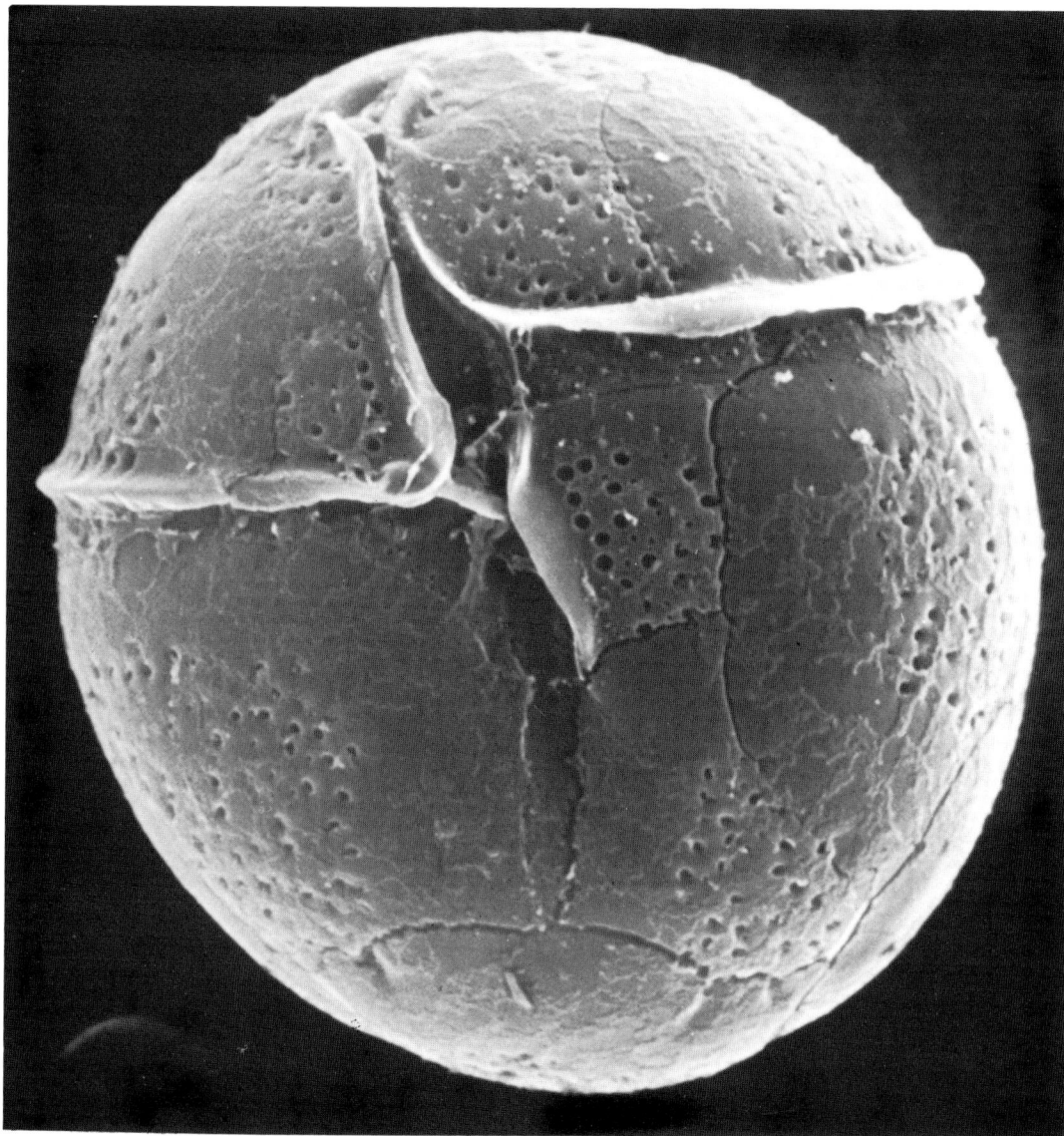

Palaeophalocroma unicinctum Schiller
A small oceanic dinoflagellate which is easily overlooked but can be recognized by its
smooth ovoid shape and the girdle which has a list only on the anterior side. Note the lack
of any sutural ridges or surface ornamentation.
N. Atlantic. 35 μm l, 30 μm w.

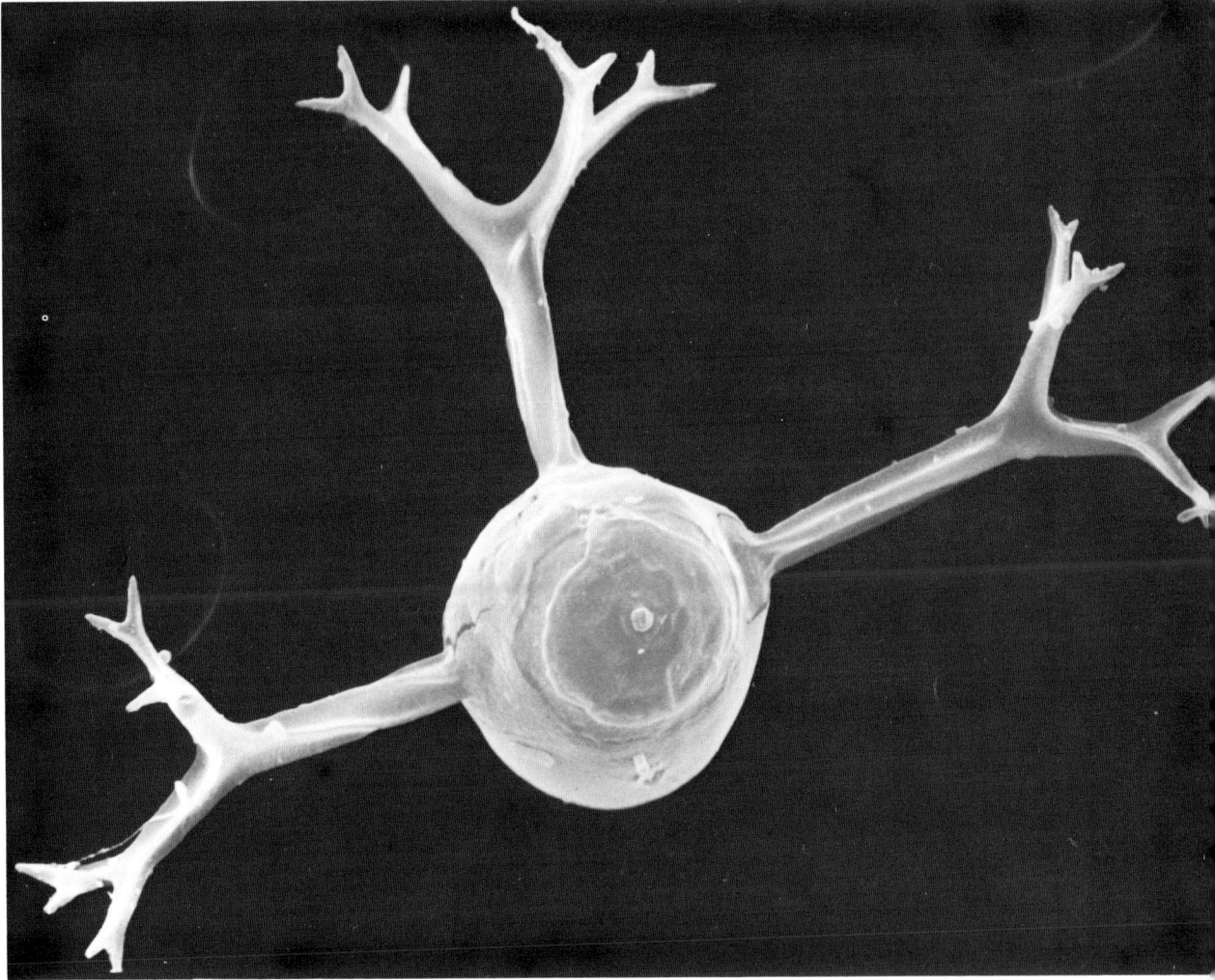

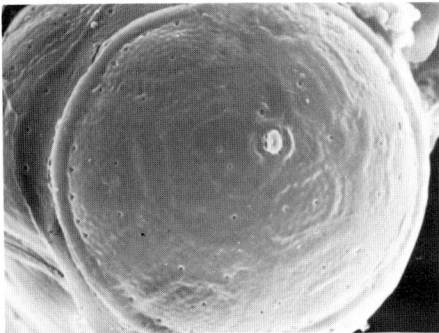

Cladopyxis brachiolata Stein
A rather rare oceanic dinoflagellate with
amazingly long bifid projections. Some
workers regard it as a cyst since the thecal plates are indistinct (small picture)
and the projections are somewhat similar
to those found on many gonyaulacoid
cysts.
E. Atlantic. Cell body 15 μm d, total
width 60 μm.

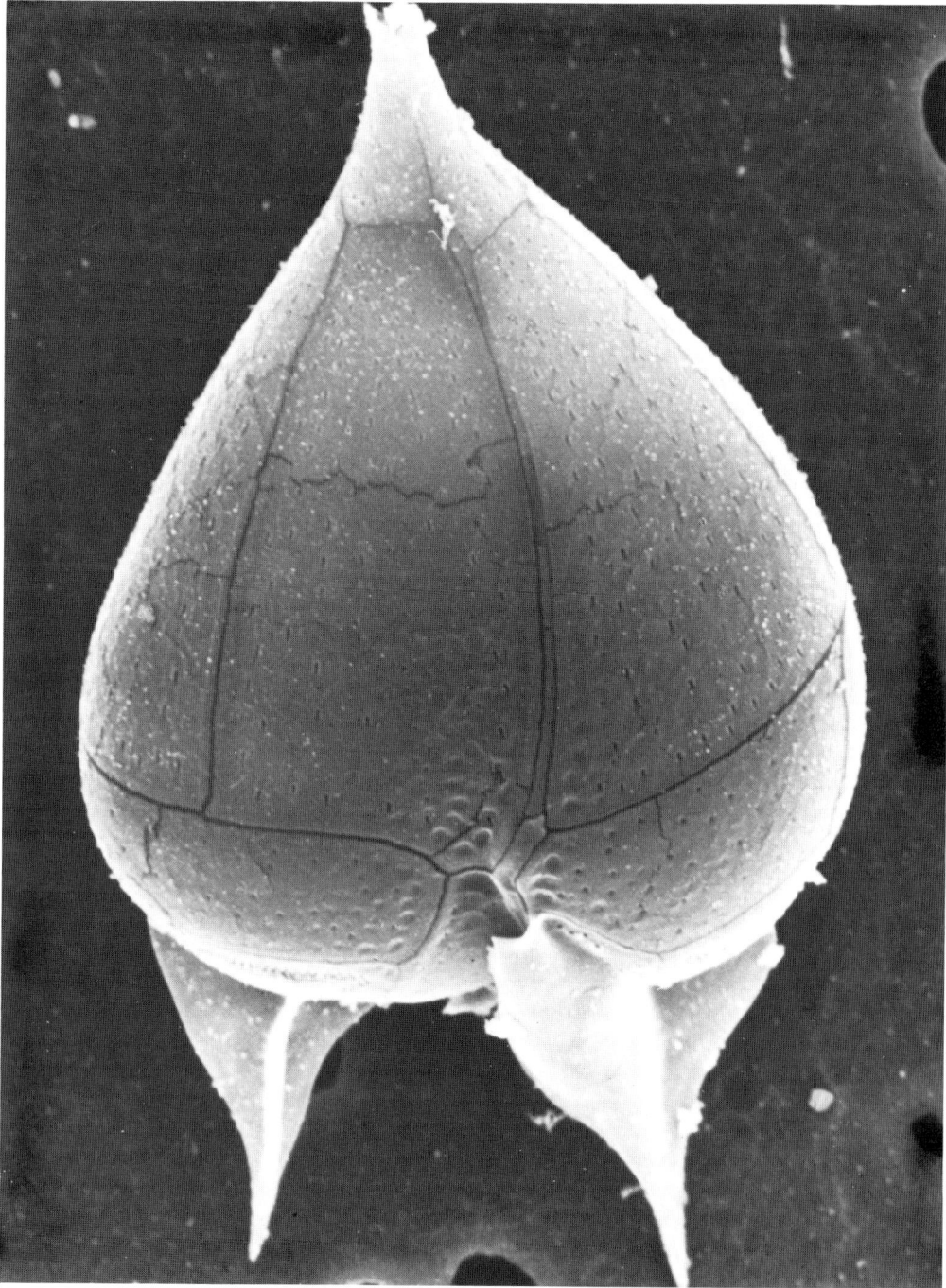

Podolampas bipes Stein

A rather unusual peridinioid dinoflagellate in that there is no obvious girdle and the surface of the cell is smooth apart from the very well developed antapical horns. The first apical plate is the narrow strip running from the sulcus to the apical horn. Members of this genus are found in warmer waters.

E. Mediterranean. 90 μm l, 55 μm w.

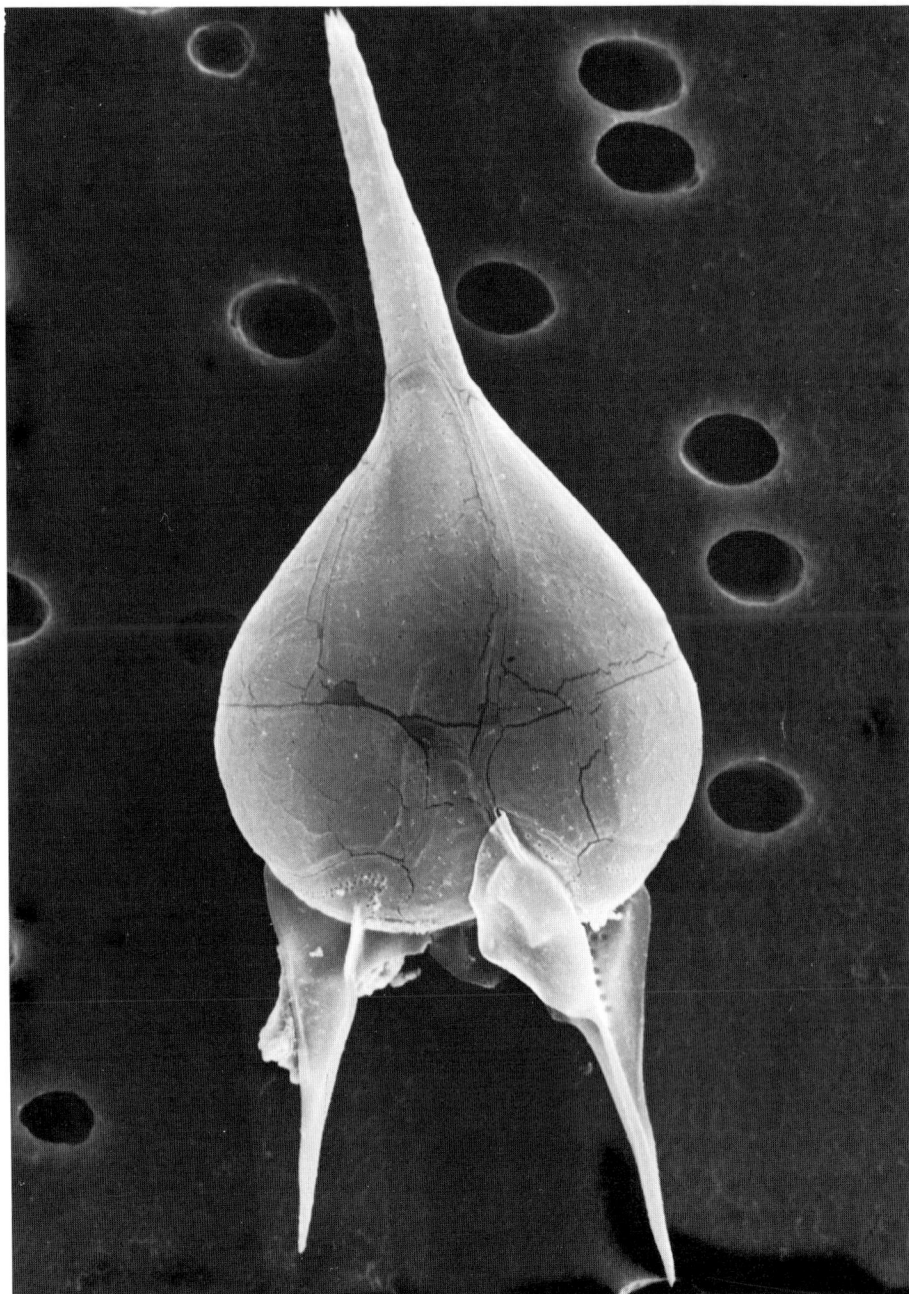

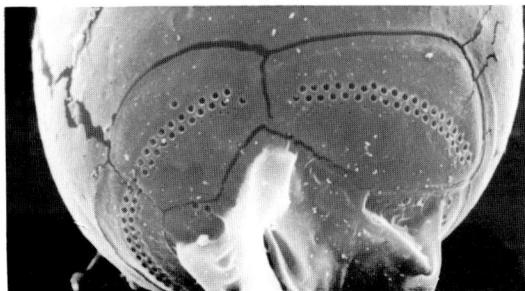

Podolampas elegans Schütt
This large dinoflagellate differs from *P. bipes* in the proportions of the cell, the apical and antapical horns being noticeably larger. The small picture shows a view of some of the hypothecal plates in *P. palmipes* with their characteristic arrangement of pores.
E. Atlantic. 120 μm l, 50 μm w.

118

Index to Species